Springer Series in Advanced Microelectronics

Volume 41

For further volumes:
www.springer.com/series/4076

The Springer Series in Advanced Microelectronics provides systematic information on all the topics relevant for the design, processing, and manufacturing of microelectronic devices. The books, each prepared by leading researchers or engineers in their fields, cover the basic and advanced aspects of topics such as wafer processing, materials, device design, device technologies, circuit design, VLSI implementation, and subsystem technology. The series forms a bridge between physics and engineering and the volumes will appeal to practicing engineers as well as research scientists.

Martin Wirnshofer

Variation-Aware Adaptive Voltage Scaling for Digital CMOS Circuits

Springer

Martin Wirnshofer
Electrical Engineering
Technische Universität München
Munich, Germany

ISSN 1437-0387 Springer Series in Advanced Microelectronics
ISBN 978-94-017-8367-5 ISBN 978-94-007-6196-4 (eBook)
DOI 10.1007/978-94-007-6196-4
Springer Dordrecht Heidelberg New York London

Softcover reprint of the hardcover 1st edition 2013

Printed on acid-free paper

Springer is part of Springer Science+Business Media (www.springer.com)

Preface

Increasing performance demands in today's integrated circuits (ICs), together with limited energy budgets, force IC designers to think about new ways of saving power. One innovative way of doing so is presented in this work. The newly developed Adaptive Voltage Scaling (AVS) scheme tunes the supply voltage of digital circuits according to the present Process, Voltage and Temperature variations as well as Aging (PVTA). The key components of the proposed approach are in-situ delay monitors (Pre-Error flip-flops), detecting late but still non-erroneous signal transitions (pre-errors). Based on the measured pre-error rate, the voltage is adjusted with a low-overhead control unit connected to the on-chip voltage regulator. This way power consumption is optimized, by exploiting unused timing margin, produced by state-of-the-art worst-case designs.

The presented Pre-Error AVS scheme adapts the voltage during normal circuit operation (on-line) and hence no test intervals, nor interaction with higher system levels is required. To describe the statistics of the closed-loop voltage control, a Markov chain model was developed, which is based on transistor level simulations. With this model, the power saving potential as well as the risk of overcritical voltage reductions can be analyzed very efficiently. Furthermore, the Markov model was extended in order to evaluate the robustness of the control-loop against global and local variations.

The elaborated approach was tested on an arithmetic and an image processing circuit, both synthesized in an industrial 65 nm low-power CMOS technology. For the two designs the Pre-Error AVS concept achieves dynamic power savings of 25 % (including all overheads) while ensuring an error rate below 1E-9. At the same time the leakage power is reduced by over 30 %.

Munich, Germany — Martin Wirnshofer

Contents

Abbreviations and Symbols

Abbreviations

AVS	Adaptive Voltage Scaling
BTI	Bias Temperature Instability
CD	Critical Dimension
CLK	Clock signal for digital logic
DC-DC	Direct Current to Direct Current
DCT	Discrete Cosine Transform
DFT	Discrete Fourier Transform
DIBL	Drain-Induced Barrier Lowering
DVFS	Dynamic Voltage Frequency Scaling
DVS	Dynamic Voltage Scaling
EDA	Electronic Design Automation
FF	Flip-Flop
FO	Fan-Out
HCI	Hot Carrier Injection
HDL	Hardware Description Language
IC	Integrated Circuit
IDCT	Inverse Discrete Cosine Transform
ITRS	International Technology Roadmap for Semiconductors
LDO	Low-Dropout
LER	Line Edge Roughness
LLN	Law of Large Numbers
MC	Monte-Carlo
MOSFET	Metal-Oxid-Semiconductor Field-Effect-Transistor
NBTI	Negative Bias Temperature Instability
OPC	Optical Proximity Correction
PBTI	Positive Bias Temperature Instability
PCM	Process Control Monitoring
PSM	Phase-Shift Masks
PVT	Process, Voltage and Temperature
PVTA	Process, Voltage, Temperature and Aging

PWCL	Pulse-Width Control Loop
RDF	Random Dopant Fluctuations
RTL	Register-Transfer Level
SoC	System-on-Chip
SPICE	Simulation Program with Integrated Circuit Emphasis
STA	Static Timing Analysis
TGMS	Transmission-Gate Master-Slave
TOX	Thickness of the gate Oxid
VHDL	Very-high-speed integrated circuit Hardware Description Language

Symbols

a	Fitting parameter for the CMOS delay model
α	Switching activity in digital circuits
$A_{\Delta k/k}$	Matching parameter for transistor current
$A_{\Delta Vt}$	Matching parameter for threshold voltage
b	Fitting parameter for the CMOS delay model
b_i	Digits of the Voltage Level control word
C_{ox}	Per unit area oxide capacitance
C_{switch}	Load capacitance of the switching gates
ΔV_{DD}	Voltage step size of the AVS regulator
$\Delta V_{di/dt}$	Voltage bounce due to di/dt noise
$\Delta V_{IR drop}$	Voltage bounce due to IR-drop
ΔP_{err}	Uncertainty of the timing error rate
f	Clock frequency
i	Time-varying current
I_D	Drain Current
I_{leak}	Leakage/static current
I_{off}	Off-state current of MOSFETs
I_{on}	On-state current of MOSFETs
k	Current factor of MOSFETs
L	Transistor length
$L_{parasitic}$	Parasitic inductance
μ	Carrier mobility
N	Number of clock cycles per observation interval
$n_{limit\downarrow}$	Lower threshold for the pre-error count
$n_{limit\uparrow}$	Upper threshold for the pre-error count
n_{pre}	Pre-error count during an observation interval
$\overline{n_{pre}}$	Mean pre-error count during an observation interval
$\boldsymbol{P}$	Markov matrix
P_{err}	Timing error rate
$P_{err,nom}$	Nominal timing error rate
P_{dyn}	Dynamic power consumption of digital circuits
$\boldsymbol{\pi}$	Stationary (steady-state) probability vector
P_{leak}	Leakage/Static power consumption of digital circuits
P_{pre}	Pre-error probability
P_{total}	Total power consumption of digital circuits

$P_{V_{DD}}$	Probability for a certain voltage level V_{DD}
$P_{V_{DD}\downarrow}$	Probability that the supply voltage V_{DD} is decreased
$P_{V_{DD}\rightarrow}$	Probability that the supply voltage V_{DD} is maintained
$P_{V_{DD}\uparrow}$	Probability that the supply voltage V_{DD} is increased
R_{grid}	Resistance of the power grid
σ	Standard deviation
τ	Time constant of the voltage regulator
t_d	Path delay
$t_{d,max}$	Maximum path delay
t_{gate}	Gate delay
T	Temperature
T_{Clk}	Clock period
T_{pre}	Length of the pre-error detection window
T_{setup}	Setup time
T_{slack}	Timing slack
V_{DD}	Supply voltage
$V_{DD,nom}$	Nominal supply voltage
V_{DS}	Drain Source Voltage
V_{GS}	Gate Source Voltage
V_{ref}	Reference voltage
V_t	Threshold voltage
W	Transistor width

Chapter 1
Introduction

Driven by Moore's Law [1], the transistor count per microchip has doubled approximately every two years for the last decades. In combination with the increase in transistor speed, the performance and functionality in solid-state circuits has improved continuously. On the other hand, this growth has made power consumption a major design criterion in digital circuits [2]. Especially in portable devices like smart-phones or tablets, the demand for increasing performance [3] with the constraints of a very limited power source makes power efficient circuits indispensable. Considering these facts, the term *performance per watt* is gaining importance for design companies, device manufacturers as well as customers.

Most of today's digital circuits are semi-custom designs, composed of standard cells. When developing digital systems, the semi-custom design flow starts with describing the logic blocks on register transfer level (RTL). This is done with hardware description languages (HDL) like Verilog or VHDL. Afterward, the logic blocks are synthesized into hardware by using electronic design automation (EDA) tools. This means, a combination of logic gates is selected from the standard cell library (containing all available logic gates) to establish the desired functionality. This set of logic gates is adjusted iteratively during the synthesis process to optimize power and area consumption. At the same time it is important that correct timing is ensured under process, voltage and temperature fluctuations as well as aging (PVTA). For the circuit, this means, it has to operate fast enough for the specified clock frequency.

To fulfill these requirements under all circumstances, the worst-case scenario is usually considered in state-of-the-art digital designs. This worst-case guard-banding approach prepares for the worst-case combination of PVTA variations by adding sufficient voltage safety margins to the nominal minimum voltage (Fig. 1.1). Note that the nominal minimum voltage is the required supply voltage if no PVTA variations occur (nominal operating condition).

Even though worst-case guard-banding guarantees proper operation in extremely suboptimal scenarios, it leads to excessively large safety margins in the standard operating conditions. Chapter 2 illustrates the different sources of variation and

M. Wirnshofer, *Variation-Aware Adaptive Voltage Scaling for Digital CMOS Circuits*, Springer Series in Advanced Microelectronics 41, DOI 10.1007/978-94-007-6196-4_1,

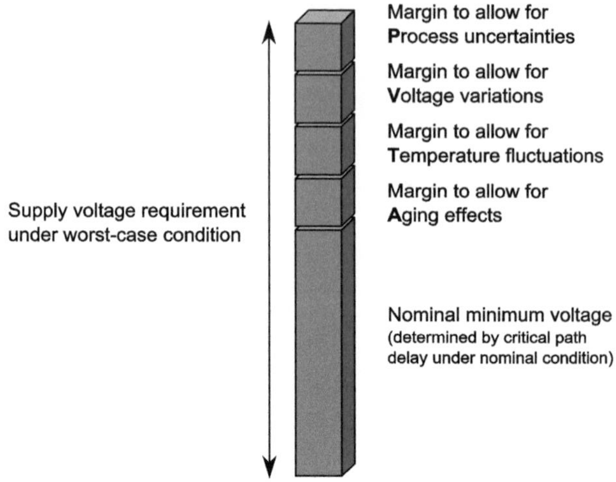

Fig. 1.1 In the worst-case guard-banding approach, voltage safety margin is introduced to account for process, voltage and temperature variations as well as aging (PVTA)

explains how shrinking technology nodes result in ever increasing safety margins, which in return deteriorate the circuit's power efficiency.

Considering the growing guard-bands in recent technologies, it is desirable to cut unused voltage margins in non worst-case situations. This way, significant power savings can be achieved as the dynamic power of digital circuits decreases quadratically with the supply voltage and the leakage power reduces even exponentially. On the other hand, reducing the supply voltage increases the path delays and may even result in timing errors.

To exploit unused safety margins, but avoid over-aggressive voltage scaling, the supply voltage has to be adjusted in a closed-loop control. This adjustment, according to the actual operating condition of the chip, is referred to as adaptive voltage scaling (AVS). To reach the necessary real-time feedback of the chip's operating condition, several approaches were proposed, which are presented in the following.

Ring-oscillators [4] and more sophisticated delay lines [5, 6] aim to replicate the timing of critical paths and are well suited for monitoring global variations on the chip. However, local (within-die) fluctuations affect the timing of these replica circuits differently than the real circuit causing again considerable safety margins. Tuning the replica circuits during test improves their tracking but adds test time and complexity [6] and therefore increases the cost of chip manufacturing.

The Razor approach [7–9] aims to monitor the timing within the real circuit, i.e. *in-situ*. Special latches [7, 8] or flip-flops [9] are used to detect timing errors and the voltage adaptation is based on the monitored error frequency. In the Razor approach, errors have to be accepted and micro-architectural recovery circuits are required, that repeat single computations after malfunction

Chapter 3 will discuss the above mentioned approaches in detail and show that monitoring the timing in-situ is a promising approach. The Razor concept however suffers from inherent problems, like the enormous complexity for the necessary recovery mechanisms. This increases design time and cost dramatically. Further problems are the lack of real-time capability and the minimum delay constraint, which is similar to the hold-time constraint in latch-based designs [10, 11].

To avoid these problems, a different in-situ monitoring approach, called Pre-Error Adaptive Voltage Scaling (AVS) is proposed in this work. In contrast to the error detection of the Razor concept, this approach is based on error prediction. Therefore, enhanced flip-flops are used that emit warnings (pre-errors) when late but still non-erroneous data transitions occur. These Pre-Error flip-flops are placed at the end of critical paths to observe the circuit's timing. The pre-error rate is an indicator for the circuit speed and is used for on-line variation and aging aware voltage adaptation. For zero or small pre-error rates, the voltage is reduced, whereas an increased pre-error rate triggers a voltage increment. In case of a moderate pre-error rate, the voltage is maintained.

Chapter 4 describes the sophisticated working principle of the Pre-Error AVS scheme, which was developed and refined during this work. The chapter also gives insight into the overall AVS control-loop, comprising the digital CMOS circuit, the in-situ delay monitors, the newly developed AVS control unit and the on-chip voltage regulator.

The Pre-Error flip-flops, acting as in-situ delay monitors, are the key part of the control-loop. They observe the timing of the circuit by detecting late data transitions. Note that a transition is late, when it occurs during the pre-error detection window, i.e. a defined time interval before the triggering edge of the clock. Hence, it is important for the AVS scheme to implement Pre-Error flip-flops with a detection window as accurate and robust as possible. During this work, existing implementations were analyzed closely and new designs for the Pre-Error flip-flop were elaborated. Chapter 5 summarizes the results and compares the implementations with respect to design effort, robustness, as well as area and power consumption.

As the Pre-Error AVS system adapts the supply voltage *on-line*, i.e. during normal circuit operation, the voltage adaptation depends on the sequence of applied input patterns and can be seen as a stochastic process. In order to describe the statistical feature of the closed-loop control, a Markov model was developed from scratch during this work, which is explained in Chap. 6. The probability data, used as input for the Markov model, were obtained by extensive transistor level simulations. Therefore, the Markov model describes the actual behavior of the system very precisely.

Chapter 7 describes how the mentioned Markov model is utilized to evaluate the power saving potential of the Pre-Error AVS approach. As test vehicle, a 16-bit multiplier circuit is used. It was synthesized in an industrial design flow for a 65 nm low-power CMOS technology. The power analysis also includes the power overhead introduced by the AVS circuitry (Pre-Error flip-flops and AVS control unit). Furthermore, the reliability of the Pre-Error AVS approach in presence of local and global variations is analyzed. It is shown, that the Markov model can simulate variations of the Pre-Error flip-flops as well as variations of the digital circuit itself very fast and accurate. With the Markov model it is also possible to estimate the risk of critical voltage reductions due to the randomness of the applied input pattern. That makes the Markov model a very powerful tool for analyzing the power efficiency as well as the reliability of Pre-Error AVS systems.

To research the applicability of the Pre-Error AVS approach to a more complex system, a discrete cosine transform (DCT) circuit was synthesized as another test object besides the multiplier circuit. As the DCT is used in current image and video coding standards, this test represents a highly relevant example. It is shown how the user can adjust the risk of critical voltage reductions by changing the AVS settings. For high image quality the settings can be chosen conservatively. For larger power savings, the reliability (image quality) can be traded, if a higher error rate is acceptable. That way the circuit can be operated at peak efficiency for the specific application.

After showing the feasibility of the proposed Pre-Error AVS approach, Chap. 8 finally summarizes the main results and gives an outlook on future trends of this technique.

Chapter 2
Sources of Variation

Variations in process, supply voltage and temperature (PVT) have always been an issue in Integrated Circuit (IC) Design. In digital circuits, PVT fluctuations affect the switching speed of the transistors and thus the timing of the logic. To guarantee fault-free operation for a specified clock frequency, IC designers have to quantify these uncertainties and account for them adequately. This is typically done by guard-banding, i.e. adding sufficient voltage safety margin to ensure proper working even under worst-case condition.

At recent technology nodes, transistor characteristics are more and more influenced also by aging effects. These wear-out effects, namely hot carrier injection (HCI) and bias temperature instability (BTI), degrade the drive current of transistors during use. Hence, further safety margin has to be added, dependent on the specified lifetime of a product.

The following four sections will give an overview of process, voltage and temperature variations as well as aging (PVTA). The necessary fundamentals are briefly explained and the impact on circuit-level timing is discussed.

2.1 Process Variations

As stated in the 2011 International Technology Roadmap for Semiconductors (ITRS) [12], "One of the key problems that designers face due to further shrinking of feature sizes is the increasing variability of design-related parameters, resulting either from variations of fabrication parameters or from the intrinsic atomistic nature which affects, e.g., channel doping." The sources for the stated device variability are treated in the following. Afterward, the impact of process variations on digital circuits is discussed.

M. Wirnshofer, *Variation-Aware Adaptive Voltage Scaling for Digital CMOS Circuits*, Springer Series in Advanced Microelectronics 41, DOI 10.1007/978-94-007-6196-4_2,

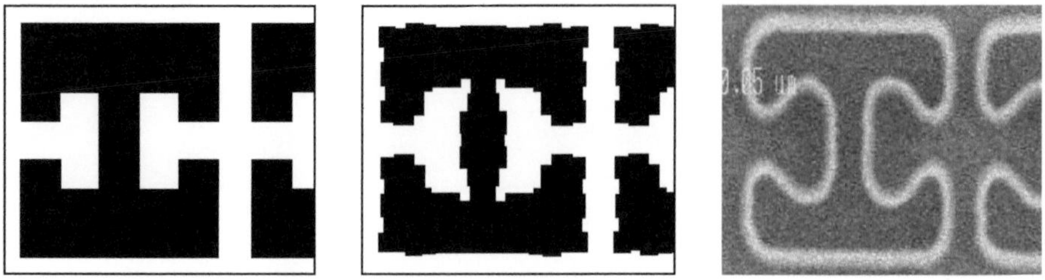

(a) Drawn structure (b) Add OPC features (c) Printed on wafer

Fig. 2.1 Diffraction causes variations in width and length of transistors. Optical proximity correction (OPC) is used to attenuate these effects [14]

2.1.1 Imperfections of the Manufacturing Process

2.1.1.1 Variations in Critical Dimensions

The wavelength of light, used for the lithography process, remained at $\lambda = 193$ nm since the 130 nm node [13]. Thus, it is exceeding more and more the minimum feature sizes, i.e. critical dimensions (CD), of transistors and the device structure gets increasingly blurred due to diffraction, see Fig. 2.1(c). Chip manufacturer try to reduce this effect by applying immersion lithography, phase-shift masks (PSM) and optical proximity correction (OPC), see Fig. 2.1(b). However, the relative fluctuations in transistor dimensions are rather increasing, as it is very challenging to reduce the absolute deviations in the same way as the shrinking feature sizes.

Variations of the transistor's width W and length L directly affect its drive current, which is proportional to W/L. Channel length fluctuations also change the threshold voltage V_t (roll-off effect) and this way additionally shift the drive current. Note that the channel length also varies randomly across the width of the transistor. The so-called line edge roughness (LER) is caused by statistical variations of the photon count or imperfections during photoresist removal [15]. LER increases the I_{off}/I_{on} current ratio for short channel devices used in digital circuits [16].

2.1.1.2 Random Dopant Fluctuation

The channel region of a transistor is doped with impurity atoms. These atoms are randomly placed into the channel—by techniques like dopant implantation—leading to statistical variations in the actual number of implanted impurities. Such a change of the carrier concentration shifts the threshold voltage and thus the drive strength of the transistor. In older technologies, with thousands of dopant atoms per channel region, an absolute deviation by several atoms was negligible. In recent technologies however the nominal number of impurities is only in the range of tens, see Fig. 2.2, leading to increased mismatch due to random dopant fluctuation (RDF).

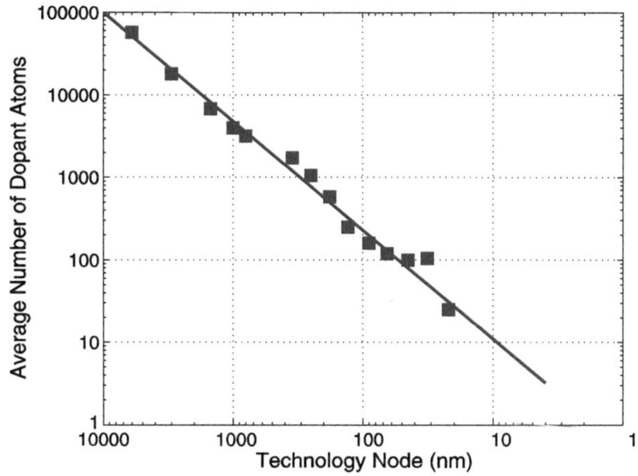

Fig. 2.2 Number of dopant atoms per channel region over technology nodes [17]

2.1.1.3 Variation of the Gate Oxide Thickness

The gate oxide can be grown with an absolute accuracy of 1–2 inter atomic layers. In former technologies with an oxide thickness (TOX) of tens of inter atomic layers and large gates, TOX induced V_t variations were almost negligible. However, in technologies below 30 nm with oxide thicknesses between 1–3 nm (approx. 5–15 inter atomic spacings), TOX variations can contribute to the threshold voltage uncertainty as much as RDF [18].

2.1.2 Global and Local Process Variations

Usually, process variations are categorized into global and local variations. For global variations device parameters, such as oxide thickness or dopant concentrations, change equally for all transistors. Wafer-to-wafer or lot-to-lot variations fall into that category. In contrast, for local variations—also known as mismatch or random uncorrelated variations—each transistor is affected differently.

In other words, variations are distinguished by their spatial correlation distance. For local variations there is no correlation, whereas for global variations the correlation distance is very large. Note that there are also variation effects in between, leading to die-to-die and within-die variations. However, to cope with process fluctuations, it is still common practice to categorize them into global and local variations [19].

2.1.3 Impact of Process Variations on Digital Circuits

Variations of physical parameters (e.g. oxide thickness or doping concentration) lead to variations of electrical parameters, like threshold voltage or gate capacitance. In turn, this affects the performance of digital circuits as it changes gate delays or leakage currents.

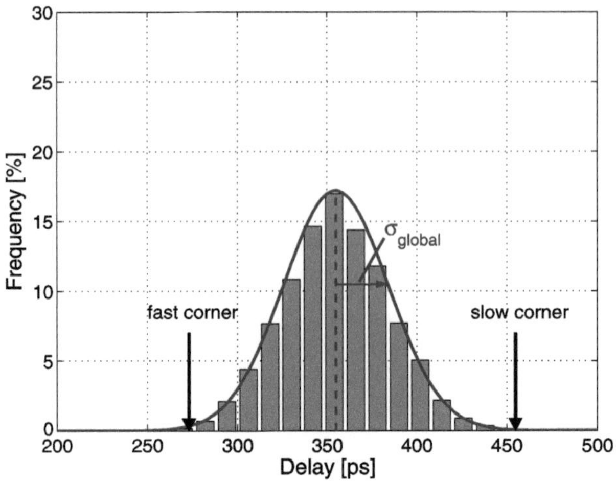

Fig. 2.3 Effect of global variations on the path delay. The depicted histogram results from MC simulations. The *arrows* represent the simulation results for fast and slow corner

2.1.3.1 Global Variations

By performing Monte-Carlo (MC) simulations on transistor level, the effect of global variations on the circuit behavior can be explored. Probability distributions for the varying device parameters serve as input for MC simulations. The distributions are based on measured statistics of the manufactured transistors. Therefore, IC manufacturers add Process Control Monitoring (PCM) structures on the scribe-line of the wafers.

The histogram in Fig. 2.3 illustrates the delay of an inverter chain in 65 nm technology under global variations. The delay values of 2000 MC-runs are divided into bins of 12 ps width. As MC simulations are computationally intensive and hence hardly feasible for large designs, corner simulations are typically used to evaluate the impact of global variations.

The corners represent extreme cases, where the devices ultimately diverge from their nominal characteristics. For the fast corner, all process fluctuations increase the drive current of a transistor, leading to maximum speed. At slow corner, a device is ultimately slowed down by the process variations. The simulation results for fast and slow corner are also drawn into Fig. 2.3.

Fast and slow corner are typically defined by multiples of σ_{global} (e.g. three or six sigma) or founded on measured performance statistics of so-called split lots. For split lots, the manufacturing process (doping concentration, oxide thickness etc.) is intentionally biased to obtain extremely slow and fast chips. Besides fast and slow corner, also cross corners exist with maximum p-FET and minimum n-FET speed and vice versa. Cross corners are often critical in analog circuits, but are of minor importance in digital designs.

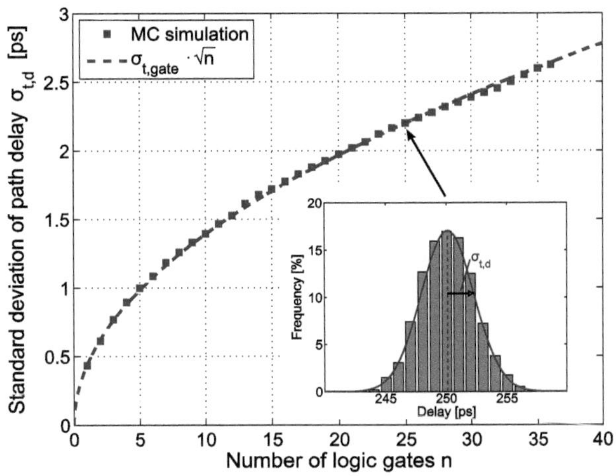

Fig. 2.4 Effect of local process variations on the path delay

2.1.3.2 Local Variations

Local variations, which are by nature uncorrelated, are increasing due to the scaling in CMOS technologies. This comes from the fact that with decreasing transistor dimensions, the standard deviation of threshold voltage $\sigma_{\Delta Vt}$ and current factor $\sigma_{\Delta k/k}$ $(k = \mu C_{ox} W/L)$ are increasing, since they are proportional to the inverse of the square root of the active device area [20]

$$\sigma_{\Delta Vt} = \frac{A_{\Delta Vt}}{\sqrt{WL}} \qquad \sigma_{\Delta k/k} = \frac{A_{\Delta k/k}}{\sqrt{WL}} \tag{2.1}$$

where $A_{\Delta Vt}$ and $A_{\Delta k/k}$ are matching parameters characterizing a particular process and manufacturing line.

The impact of local variations on the path delay can be described by a stochastic model of n equal logic gates with a switching time (rise or fall time) of t_{gate} and the corresponding standard deviation of $\sigma_{t,gate}$. The path delay $t_d = n \cdot t_{gate}$ increases linearly with n, whereas the standard deviation of the path delay $\sigma_{t,d}$ increases only proportional to the square root of n and reads as follows[1]

$$\sigma_{t,d} = \sqrt{n} \cdot \sigma_{t,gate} \tag{2.2}$$

Thus, the relative variation of the path delay $\sigma_{t,d}/t_d$ decreases proportionally to the inverse square root of n

$$\frac{\sigma_{t,d}}{t_d} = \frac{1}{\sqrt{n}} \cdot \frac{\sigma_{t,gate}}{t_{gate}} \tag{2.3}$$

---

[1] The derivation for $\sigma_{t,d}$ can be found in the Appendix.

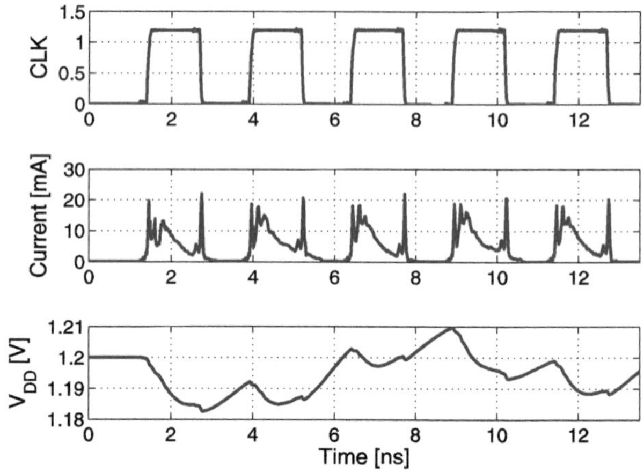

Fig. 2.5 Clock signal, current profile and resulting supply voltage of a 32-bit multiplier circuit. Close to the clock edges, a high current flow can be seen due to the switching activity of the circuit and flip-flops. The combination of IR-drop and di/dt noise in turn leads to voltage drops and—if resonance occurs—also to overshoots

It is a common pitfall, to think that the absolute uncertainty $\sigma_{t,d}$ decreases for long paths. However, only the *relative* uncertainty $\sigma_{t,d}/t_d$ decreases with the path length, whereas the absolute uncertainty $\sigma_{t,d}$ naturally increases, see Fig. 2.4.

2.2 Voltage Variations

2.2.1 Origin of Voltage Fluctuations

Supply voltage fluctuations are mainly caused by IR drop and di/dt noise. IR drop is caused by the current flow over the parasitic resistance of the power grid, whereas di/dt noise is due to the parasitic inductance in combination with capacitance and resistance of power grid and package. These fast changing effects—also called power noise—typically have time constants in the range of nano- to microseconds. Figure 2.5 shows an example of a supply current profile together with the resulting supply voltage sequence for a 32-bit multiplier circuit.

Considering only IR drop, the voltage bounce $\Delta V_{IR\,drop}$ is given by Ohms law

$$\Delta V_{IR\,drop} = R_{grid} \cdot i(t) \tag{2.4}$$

Additionally, the current flow through a parasitic inductance $L_{parasitic}$ causes the voltage bounce

$$\Delta V_{di/dt} = L_{parasitic} \cdot \frac{\mathrm{d}i}{\mathrm{d}t} \tag{2.5}$$

The superposition of both effects can not only lead to voltage drops, but also voltage overshoots, as can be seen in Fig. 2.5.

Besides fast changing power noise effects, also offsets in the voltage regulator can lead to deviations from the nominal supply voltage. Offsets can either result from inaccuracies of the voltage regulator itself, e.g. a DC-DC or LDO regulator, or from the voltage reference circuit, often implemented as bandgap circuit.

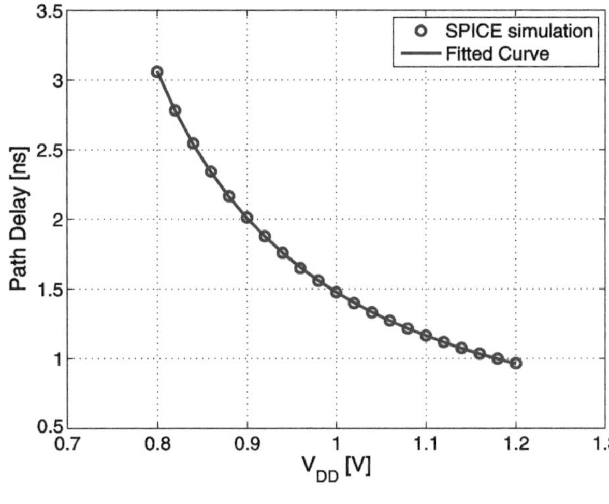

Fig. 2.6 Relation between supply voltage and logic delay, illustrated by the most critical path of a multiplier circuit in 65 nm technology. The simulation results can be accurately fitted by Eq. 2.7

2.2.2 Impact of Voltage Changes on the Path Delay

The delay of a CMOS logic gate (e.g. an inverter or a NAND-gate) can be approximated by the following equation [21]

$$t_{gate} \propto \frac{V_{DD}}{b(V_{DD} - V_t)^a} \tag{2.6}$$

where a and b are gate specific fitting parameters and V_t is an effective threshold voltage. The delay of a complete path is obtained by summing up the single gate delays along this path. Consequently, the output delay t_d can be approximated by

$$t_d \propto \frac{V_{DD}}{b'(V_{DD} - V_t')^{a'}} \tag{2.7}$$

with path-specific parameters a', b' and V_t'. Figure 2.6 shows the simulated path delay of a multiplier circuit for different supply voltages. By applying the appropriate input pattern, the most critical path was sensitized in the simulation. The solid line demonstrates the fit by Eq. 2.7.

2.3 Temperature Variations

Dependent on the thermal conductivity, the dissipated power affects the temperature of a chip. Power dissipation hence leads to global temperature variations as well as local fluctuations in regions of high-activity, so called hot-spots. Additionally, ambient temperature changes lead to global shifts in chip temperature.

Temperature fluctuations typically have time constants in the range of milliseconds to seconds [22]. An increase in temperature typically causes a circuit to slow down due to reduced carrier mobility and increased interconnect resistance, see

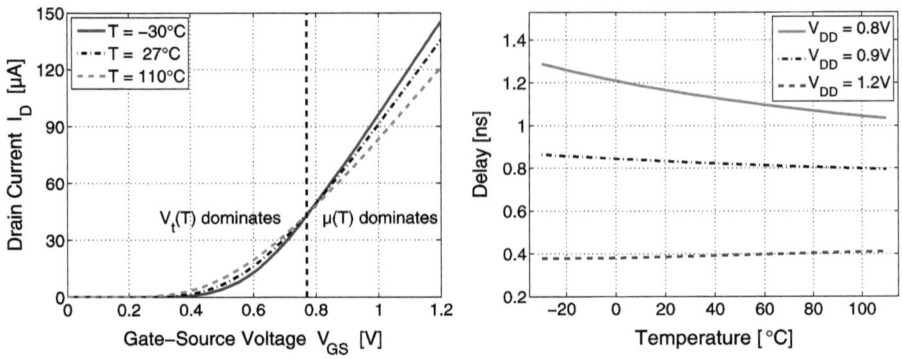

(a) Transfer characteristic of a MOSFET (b) Path delay for different voltages

Fig. 2.7 Influence of temperature on device characteristic and path delay

Fig. 2.7(b) ($V_{DD} = 1.2$ V) and [23]. However, for low V_{DD} the circuit is operated in temperature inversion. Here, the effect of decreasing threshold voltage with temperature exceeds the mobility degradation, see Fig. 2.7(a). Consequently, the circuit exhibits an inverted temperature characteristic, as it speeds up with increased temperature and vice versa, see Fig. 2.7(b) ($V_{DD} = 0.8$ V and $V_{DD} = 0.9$ V).

2.4 Aging

Due to increasing electrical fields and new materials, transistor wear-out is of increasing concern in recent technologies. BTI- as well as HCI-effects degrade the speed of transistors during their lifetime and demand for additional safety margin.

2.4.1 Hot Carrier Injection (HCI)

Hot Carrier Injection mainly occurs during switching of logic gates. Carriers are accelerated in the lateral field under the oxide and gain sufficient kinetic energy to be injected into the gate dielectric. This effect is illustrated for an n-MOSFET in Fig. 2.8. The trapped charge increases the threshold voltage of the device and reduces its current drivability. Recent research on HCI is presented in [24].

2.4.2 Bias Temperature Instability (BTI)

BTI results from high vertical fields and thus mainly occurs when a transistor is operated in triode mode (linear region), i.e. high V_{GS} and low V_{DS} is applied. Figure 2.9 shows charge trapping by BTI for an n-FET, referred to as Positive Bias

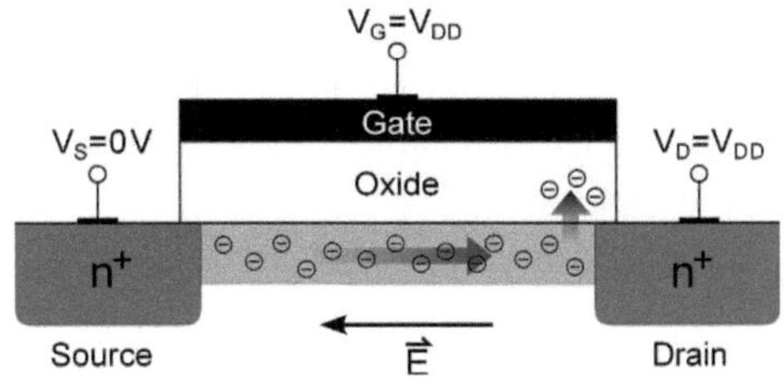

Fig. 2.8 Hot Carrier Injection (HCI) for an n-MOSFET

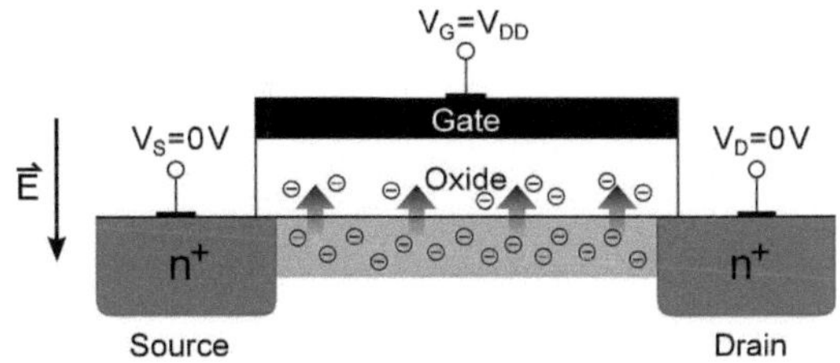

Fig. 2.9 Bias Temperature Instability (BTI) for an n-MOSFET

Temperature Instability (PBTI). Accordingly, the term NBTI is used in the case of a p-FET. PBTI and NBTI increase the threshold voltage of the device and slow down the switching speed [25].

Note that BTI-aging is caused by charge trapping and detrapping with a wide range of capture and emission times [26]. Therefore, small V_t-shifts can be observed already after very short stress times down to microseconds [27]. However, due to the distribution of the capture and emission time constants, considerable V_t-shifts arise only after days, weeks or even years. More details on BTI can be found in [28–31].

2.5 Summary

Figure 2.10 categorizes the discussed variations according to their time constants. As the dynamics of variation effects depend on various circumstances, the shown classification is rather qualitatively.

Process variations occur during fabrication and lead to fixed changes in device parameters. During operation, device characteristics can, however, still be affected by aging induced wear-out. As mentioned before, the trapping events leading to device aging have a wide variety of time constants. However, significant degradation of CMOS logic delays mainly develops in the long run. The temperature, also influencing circuit speed, changes rather slowly, whereas power noise has normally very short time constants.

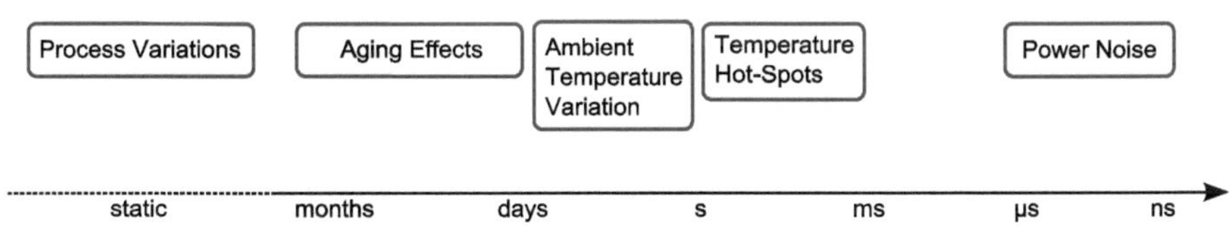

Fig. 2.10 Temporal classification of variations

The shorter the time constant of a variation effect, the more challenging becomes the voltage adaptation to it. In the following chapter, it is explained how Pre-Error AVS effectively handles slow as well as fast changing effects compared to state-of-the-art approaches.

Chapter 3
Related Work

As discussed in the previous chapter, process, voltage and temperature variations as well as aging significantly affect the timing of digital circuits. To cope with these uncertainties, worst-case guard-banding is still the most common design approach. As the worst-case is very rare however, in most cases power or performance is wasted by this approach.

By scaling the operating voltage, energy efficiency can be increased. Tuning the supply voltage dependent on PVTA variations is referred to as adaptive voltage scaling (AVS). The term AVS is often interchanged with dynamic voltage scaling (DVS), but note that DVS considers only varying workloads and does not adapt to PVTA variations at all. To clarify the difference between both techniques, they are explained in the following two sections.

3.1 Dynamic Voltage Scaling

Dynamic Voltage Scaling (DVS) relies on pre-characterized pairs of operating frequencies and corresponding supply voltages. These pairs are stored in look-up tables and selected during run-time dependent on the workload or performance demand of a chip [32].

For commercial use, DVS was first applied in microprocessors. A well-known example is Intel's SpeedStep Technology, where the operating system can trigger a frequency/voltage change by writing the dedicated register [33]. Table 3.1 illustrates the frequency/voltage look-up table for a Pentium M Processor with DVS capability. "P-State" denotes the power state level, with "P0" defined as state with the fastest performance but in consequence also the maximum power consumption.

When the performance demand is low or when a notebook is run on battery power the frequency can be lowered and the corresponding voltage is applied at the same time. As the power dissipation increases quadratically with the applied supply voltage, a voltage reduction decreases the energy consumption significantly.

It is important to underline that for DVS the respective voltage for each frequency is characterized by worst-case guard-banding during design stage. Changes of the

M. Wirnshofer, *Variation-Aware Adaptive Voltage Scaling for Digital CMOS Circuits*,
Springer Series in Advanced Microelectronics 41, DOI 10.1007/978-94-007-6196-4_3,

Table 3.1 DVS look-up table with pre-characterized pairs of operating frequencies and corresponding supply voltages for an Intel Pentium M Processor [34]

Frequency	Voltage	P-State
1.6 GHz	1.484 V	P0
1.4 GHz	1.420 V	P1
1.2 GHz	1.276 V	P2
1.0 GHz	1.164 V	P3
800 MHz	1.036 V	P4
600 MHz	0.956 V	P5

hardware's operating condition during run-time due to PVTA variations are thus not considered at all. In consequence, DVS—often also named DVFS (Dynamic Voltage Frequency Scaling)—does not improve energy efficiency in fixed frequency scenarios.

3.2 Adaptive Voltage Scaling

In contrast to DVS, Adaptive Voltage Scaling (AVS) adjusts the supply voltage according to the actual operating condition of a chip. For this closed-loop control it is thus essential to measure the circuit's operating condition, including all sorts of PVTA variations. To solve this task several approaches have been proposed and are described in the following.

3.2.1 AVS by Replica Paths

Using replica paths for voltage adaptation was first proposed in 1990 [35]. The basic idea is to mimic the timing of the most critical path by a replica circuit. Instead of measuring the speed of the real circuit, the delay of the replica path can thus be used to tune the operating voltage.

Copying the layout of the most critical path including its wire connections as well as considering all side inputs and coupling effects is rather complex. Instead, idealized paths—like an inverter chain—are typically used for emulating the critical timing, see Fig. 3.1.

While replica circuits are well-suited for monitoring global variations on the chip, still sufficient safety margin has to be included to cover the diverging timing behavior of critical and replica path over the whole voltage and temperature range. For fault-free operation, it has to be ensured that the replica path is always slower than the real circuit.

Even when implementing more sophisticated delay lines with a programmable mixture of gate and wire delays [5, 6]—instead of simple inverter chains or ring-oscillators—the necessary safety margin can only be reduced to some extent.

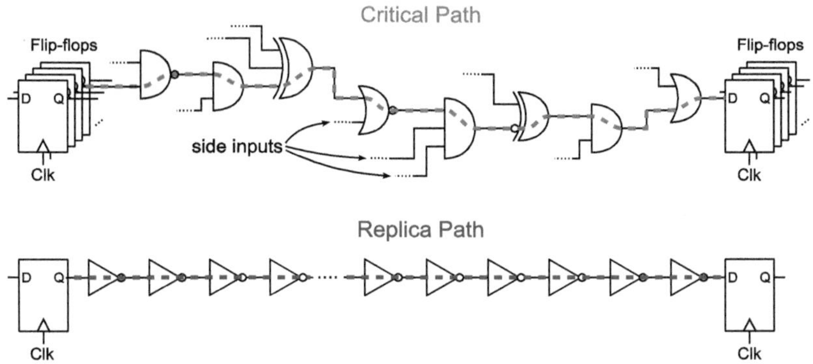

Fig. 3.1 The replica path mimics the critical timing of the digital logic

Moreover, within-die variations of process, voltage, temperature and especially aging influence the timing of replica circuits differently than the real circuit [36, 37], requiring additional safety margins. Tuning the replica circuits during test improves their tracking but adds test time and complexity [6] and therefore increases cost.

Also note that different signal propagation paths may get critical dependent on PVTA or workload variations. Thus, identifying the most critical one is challenging and it is almost impossible to rule out the existence of any other path being even more critical under some circumstance. This uncertainty again demands for a proper security margin and further decreases the efficiency of AVS by replica paths.

3.2.2 *Razor Concept*

To circumvent the imperfect tracking of replica paths, recent approaches aim to monitor the timing *in-situ*, i.e. within the real circuit. The so-called Razor approach, proposed by [10], replaces the flip-flops at the end of critical paths by Razor flip-flops, which are able to detect timing errors. For this purpose, a shadow latch is added in parallel to the regular flip-flop as depicted in Fig. 3.2.

The regular flip-flop samples the incoming data at the rising clock edge, whereas the shadow latch is in transparent mode during the entire clock-high phase. If the incoming data meets the setup time of the flip-flop, latch and flip-flop will propagate the same data to their output, see Fig. 3.2(b) (Cycle 1). For timing failures, the shadow latch still captures the valid data, while the flip-flop fails to sample the correct value (Cycle 2). Consequently, the outputs Q and Q_{shadow} differ and the comparator (XOR-gate) flags an error. The error signals of all implemented Razor flip-flops are than OR'ed together for the final restore signal. In case of an error the correct data can be restored by the multiplexer.

In the Razor approach, the voltage adaptation is based on the measured timing error rate. The voltage is set to a value, where a small, but non-zero error rate is obtained. As errors have to be accepted, architectural recovery circuits are necessary,

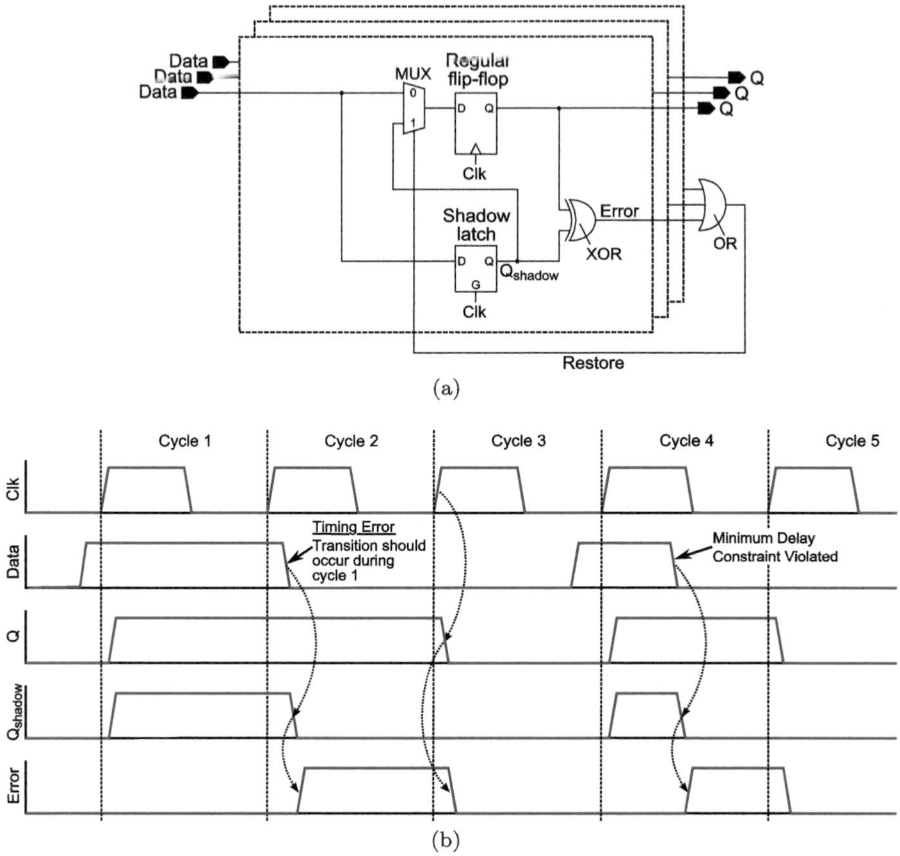

Fig. 3.2 Basic schematic of a Razor flip-flip (**a**) proposed by [10] and the corresponding timing diagram (**b**). Erroneous transitions are detected as timing errors

that repeat single computations after malfunction. For small error rates, the power overhead caused by repeating the erroneous computations is small. However, the additional complexity for the required recovery architecture is enormous. The Razor implementations for a 3-stage pipeline [38] and a subset of an ARM processor [39] use customized stall mechanisms and instruction-replays. Besides this undesirable increase in complexity, error recovery also leads to variable and unpredictable latency. Therefore, Razor AVS is not suitable for real-time applications. Particularly problematic are phases with relaxed data sequences or instructions followed by critical inputs. During the relaxed phase, the system will scale the voltage aggressively and suffer from an increased error rate/latency afterward.

Another problem of the Razor concept is its inherent minimum delay constraint [11], which is similar to the hold-time constraint in latch-based designs. As the path delay in combinatorial logic is strongly data dependent, delays at one and the same flip-flop can be critical in one clock cycle and relaxed in the following.

Since all transitions during the error detection window (high clock phase) result in an error signal, the Razor flip-flop cannot distinguish between a timing failure from the previous clock cycle and a very fast computation in the current clock cycle (false error). This is exemplified in Fig. 3.2(b) (Cycles 3 and 4). Though the transition in Cycle 4 is uncritical, it is detected as a timing error of Cycle 3.

False errors cause needless repetition of computations resulting in a waste of power and time. What is even worse, repeating these computations will again lead to an error signal and the system falls into a fatal deadlock situation.

The longer the error detection window, the more likely false errors become. On the other hand, the error detection window has to be long enough to detect all over-critical transitions. Otherwise the Razor flip-flop misses timing errors, leading to uncorrected and consequently catastrophic failures.

By tuning the duty cycle of the clock, the error detection window can be adapted.[1] Duty cycle adaptation was implemented either by a separate clock tree for the Razor flip-flops [8], which adds significant routing and area overhead, or by changing the duty cycle for the whole chip [11, 38]. Another possibility was presented in [39], where a classical 50 % duty-cycle can be distributed and the error detection window length is adapted locally in the Razor flip-flop.

For an adequate length of the error detection window, the previously explained *minimum delay violations* are unavoidable. To ensure that the delays in all Razor equipped paths fulfill the minimum delay constraint, buffers have to be inserted in some paths [8, 10, 11, 38–40]. This workaround however increases the maximum path delay and is obviously counterproductive as it limits voltage scaling. Moreover, the buffers consume extra area and power.

3.2.3 Pre-Error AVS

Monitoring the timing inside the real circuit (in-situ), as in Razor AVS is a promising approach. However, the Razor concept suffers from inherent problems, like the additional complexity for the necessary recovery mechanisms, which increases design time and cost tremendously. Further obstacles are the lack of real-time capability and the described minimum delay constraint or hold time problem.

To avoid these problems, in [41] a different in-situ monitoring approach is proposed. In contrast to the error detection of the Razor concept, in this approach, enhanced flip-flops detect late but still non-erroneous transitions (pre-errors). These flip-flops are placed at the end of critical paths to observe the circuit's timing. Figure 3.3 shows the basic structure of such a Pre-Error flip-flop and the corresponding timing diagram.

The Pre-Error flip-flop comprises the regular flip-flop and a second flip-flop in parallel. For the second flip-flop the data input is delayed by the time T_{pre}. For a

[1] Arbitrarily changing the duty cycle is of course only possible, if the negative clock edge is not used to trigger any logic events.

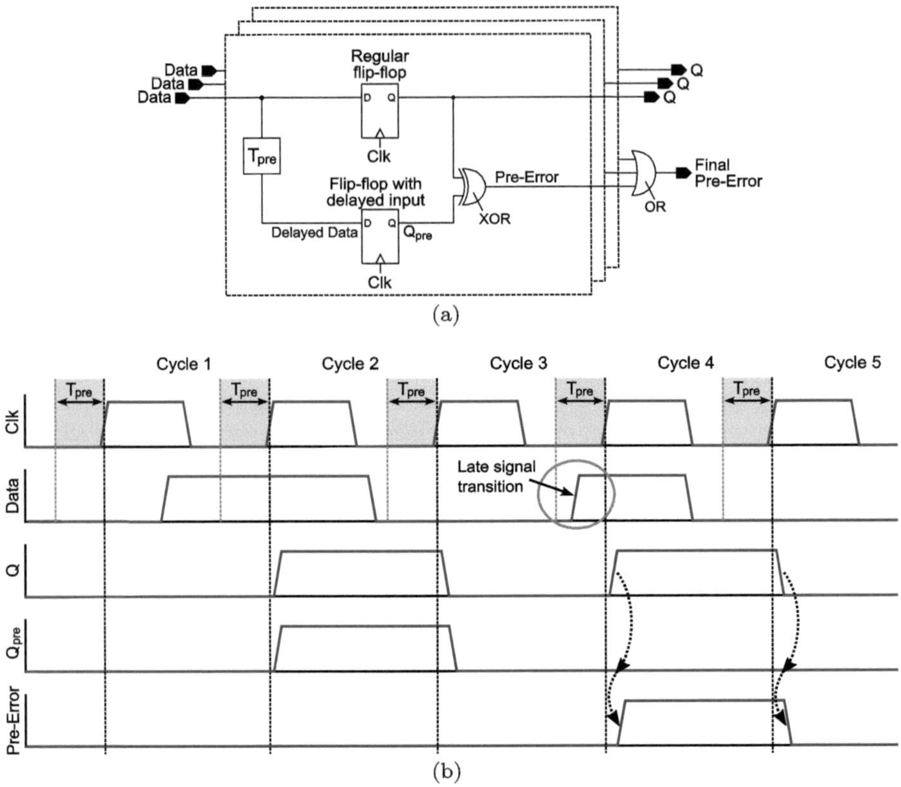

Fig. 3.3 Basic schematic of a Pre-Error flip-flip (**a**) proposed by [41] and the corresponding timing diagram (**b**). Late data transitions (*encircled*) are detected and a pre-error is flagged

timing slack larger than T_{pre} both flip-flops sample the same data, see Fig. 3.3(b) (Cycles 1, 2 and 4). For timing slacks shorter than T_{pre}, the second flip-flop fails to capture the correct value and the outputs Q and Q_{pre} differ, see Fig. 3.3(b) (Cycle 3). This difference is detected by the XOR-gate and a pre-error is flagged. All pre-error signals are then OR'ed together for the final pre-error value.

As the warnings are flagged before errors occur, no additional hardware effort and complexity for recovery circuitry is needed. In [41, 42], it was proposed to use these monitoring flip-flops in combination with test vectors stored on-chip and applied during test-sequences. Test intervals require, however, interaction with higher system levels, introduce latency and increase the overall system complexity. Moreover, finding test vectors that sensitize the most critical paths with high coverage is very challenging. During a circuit's lifetime, the critical patterns are also likely to change due to varying operating conditions and especially aging [36].

It is also suggested to use in-situ delay monitoring (pre-error detection) during normal operation [41]. No test vectors are thus needed, nor test intervals that interrupt the circuit operation. This way, the AVS system can run autonomously without need for interaction with higher system-levels. However, this on-line voltage adaptation was only touched briefly and no suggestions were made how to implement it. Therefore, it was the goal of this work to elaborate concrete ideas how to realize such a system and investigate its feasibility.

Chapter 4
Adaptive Voltage Scaling by In-situ Delay Monitoring

4.1 Principle of Operation

In the Pre-Error AVS scheme the timing information is provided by in-situ delay monitors (Pre-Error flip-flops), that detect late but still non-erroneous data transitions in critical paths. Late data transitions are defined by the pre-error detection window, i.e. a defined time interval T_{pre} before the triggering edge of the clock.

The timing of digital circuits is influenced by PVTA variations and so is the frequency of pre-errors. The pre-error rate, indicating the timing slack, is thus used to adapt the supply voltage on-the-fly/on-line, i.e. during normal circuit operation. During each N clock cycles, forming an observation interval, the number of pre-errors n_{pre} is counted and it is decided whether to change the voltage subsequently. For a pre-error count n_{pre} under a lower threshold $n_{limit\downarrow}$, the voltage is decreased by ΔV_{DD}. If the count is above the upper limit $n_{limit\uparrow}$, the voltage will be increased by ΔV_{DD}. For counts inside the limits the voltage is maintained, see Fig. 4.1.

4.2 Overall AVS Control Loop

Adding up the detected pre-errors during fixed observation intervals and defining threshold values for the pre-error count, as described above, is an innovative method for adaptive voltage scaling. It was invented [43] and refined during the work on this

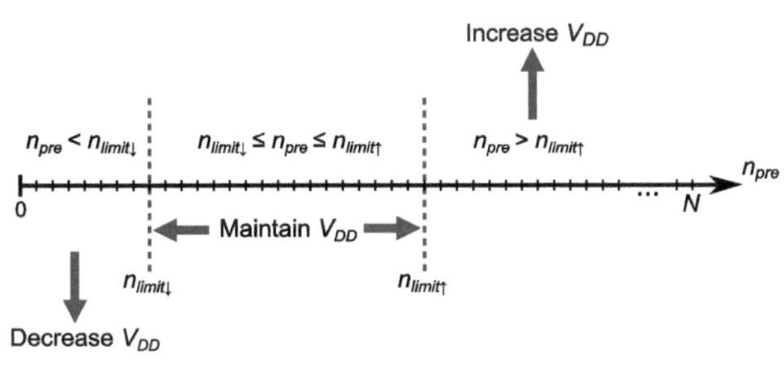

Fig. 4.1 The voltage control in the Pre-Error AVS approach is based on the pre-error count n_{pre} during observation intervals, consisting of N clock cycles

M. Wirnshofer, *Variation-Aware Adaptive Voltage Scaling for Digital CMOS Circuits*, Springer Series in Advanced Microelectronics 41, DOI 10.1007/978-94-007-6196-4_4,

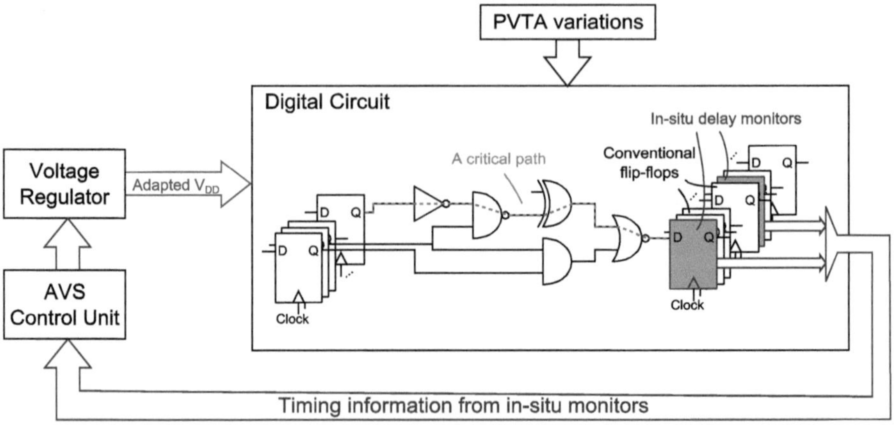

Fig. 4.2 Block diagram of the overall Pre-Error AVS control loop. The timing of the digital circuit is measured by in-situ delay monitors (Pre-Error flip-flops). Based on the obtained timing information, the AVS control unit triggers the voltage regulator to adapt the supply voltage V_{DD}

thesis [44]. In order to evaluate the feasibility and potential of the new concept, all integral parts of the Pre-Error AVS system were analyzed. The closed-loop configuration of this voltage control system is outlined in Fig. 4.2.

Under PVTA variations the operating condition of digital circuits changes. For instance, device parameters vary due to aging effects or the temperature increases during heavy workload. For AVS, however, only the timing of the logic is relevant, no matter if it is changed by transistor aging or a shift in temperature. To measure the timing, in-situ delay monitors (Pre-Error flip-flops) are placed at the end of critical paths. The AVS control unit [45] counts the detected pre-errors and triggers the on-chip voltage regulator to adjust the supply voltage V_{DD} if necessary. The AVS control unit and the voltage regulator are described more detailed in Sects. 4.2.3 and 4.2.4, respectively. Before, the role of the digital circuit itself and the in-situ delay monitors will be discussed.

4.2.1 Digital CMOS Circuit

Due to the large complexity of digital blocks in modern Systems-on-Chip (SoC), most of the digital circuits are semi-custom designs nowadays. Compared to expensive full-custom designs, the semi-custom approach saves a huge amount of development time and cost, with a moderate penalty in performance, power or area.

In semi-custom designs, the desired functionality is first described on register transfer level (RTL) with a Hardware Description Language[1] (HDL). Afterward,

[1]The two most common hardware description languages are VHDL and Verilog.

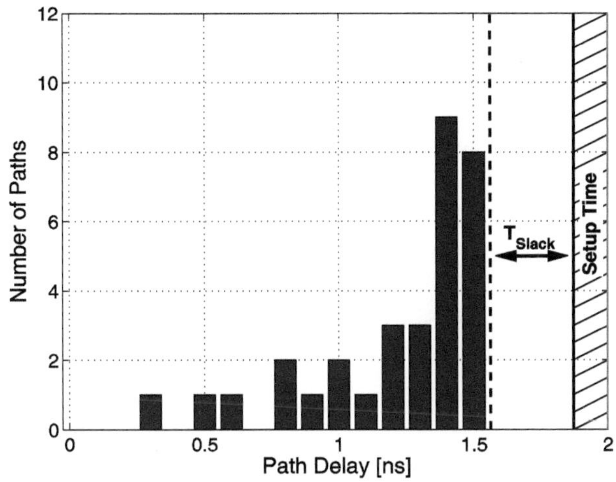

Fig. 4.3 Path delay distribution of a 16-bit multiplier circuit, synthesized in a commercial 65 nm low-power design flow with state-of-the-art EDA tools. For each output, its maximum path delay contributes to the histogram. The delays are simulated on transistor level by applying the most critical input patterns

the HDL code is synthesized into hardware by using electronic design automation (EDA) tools. These tools select a set of standard cells (logic gates) to realize the desired functionality. All available standard cells for a given process technology are full-custom designed beforehand and stored in a standard library. For a specified clock frequency, the synthesis tools try to optimize area and power consumption by choosing the best combination of standard cells. Finding the optimal selection of standard cells is done iteratively by sophisticated algorithms.

During the synthesis iterations, the slack in uncritical paths is exploited to insert slower and less power-consuming gates along these paths. However, slowing down uncritical paths is not arbitrarily possible as uncritical paths often have sub-paths in common with critical paths. Thus, balancing the timing is limited and the majority of outputs remains uncritical in typical digital designs. This is exemplified by Fig. 4.3, showing the path delay distribution of a 16-bit multiplier circuit. It was synthesized in a commercial 65 nm low-power technology with state-of-the-art EDA tools. The clock frequency was set to $f = 500$ MHz ($T_{Clk} = 2$ ns) and the synthesized circuit is comprised of 701 logic gates from the standard cell library. The total number of flip-flops is 64 (32 for the multiplicands at the input and 32 for the resulting product at the output).

For each output, its maximum path delay contributes to the histogram in Fig. 4.3. We obtained the shown distribution by applying the most critical input patterns to the synthesized circuit and simulating them with SPICE on transistor level.

Since most paths are uncritical, in Pre-Error AVS it is sufficient to equip only a small percentage of outputs with in-situ delay monitors. Thus, the power and area overhead due to insertion of Pre-Error flip-flops is reasonable.

Note that synthesis tools are based on static timing analysis (STA). This means they can only approximate the timing of the circuit and therefore have to be rather conservative (see also Sec. 6.1 on page 45 for detailed explanation). This necessary pessimism adds up in additional timing margin for the synthesized circuits.

Fig. 4.4 Delay histogram for the most critical output of the synthesized multiplier circuit, when applying random input patterns. The delay values are obtained by transistor level simulations with SPICE

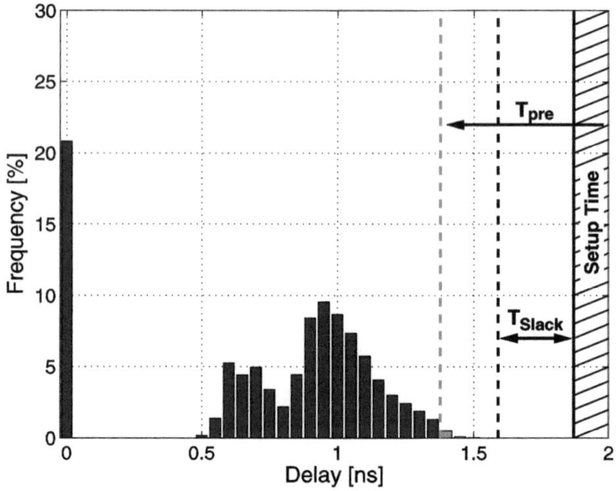

Besides different maximum arrival times for different output nodes, see Fig. 4.3, for a certain output the delay strongly varies in dependence on the applied input patterns. This is due to changing signal propagation paths inside the combinatorial logic with varying input vectors. Figure 4.4 illustrates this effect for the most critical output of the synthesized multiplier circuit. The histogram is obtained by transistor level simulations of random input patterns. As an example, a pre-error detection window length of $T_{pre} = 600$ ps is visualized by the left dashed line. With a clock period of $T_{Clk} = 2$ ns, delays longer than $t_d = T_{Clk} - T_{pre} = 1.4$ ns will result in a pre-error for this detection window.

The delay distribution and consequently the pre-error probability depend on the operating condition of the circuit. Aging, for example, shifts the distribution to the right and increases the pre-error rate. The pre-error rate is thus used as indicator for the operating condition or the remaining timing slack ($T_{slack} = T_{Clk} - T_{setup} - t_{d,max}$), respectively. Note that the occurrence of pre-errors depends on the distribution of the output delays and can be interpreted as a stochastic process. Hence, the pre-error count during a given time span can be seen as a random variable.

Regarding a single output, the pre-error probability per clock cycle is the number of output delays longer than $T_{Clk} - T_{pre}$ divided by the total number of clock cycles. Note that in our AVS system, the pre-errors of all the observed outputs are OR'ed together, resulting in the overall pre-error rate P_{pre}.

4.2.2 In-situ Delay Monitor

The Pre-Error flip-flop acting as in-situ delay monitor is of key importance for the proposed AVS approach. It is an enhanced flip-flop, that is able to observe the timing by detecting late transitions. As mentioned above, late transitions are defined by the pre-error detection window with length T_{pre}. Transitions inside the detection

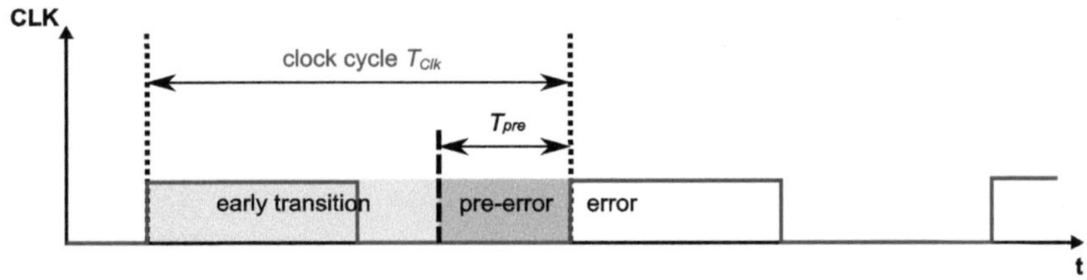

Fig. 4.5 Data transitions inside the pre-error detection window are detected as pre-errors by the in-situ delay monitor. The length of the detection window T_{pre} is a major design parameter for the Pre-Error AVS scheme

window are detected as pre-errors, whereas earlier transitions remain undetected, see Fig. 4.5. Typical values for T_{pre} are around 30 % of the clock cycle T_{Clk}. For the Pre-Error AVS scheme it is of high importance to define the detection window length as accurate and robust as possible. Details on the design of the Pre-Error flip-flop and different ways how to define the detection window are separately discussed in the next chapter.

The voltage control in the Pre-Error AVS approach is based on the pre-error count n_{pre} during observation intervals, consisting of N clock cycles. Considering the activity rate α of a circuit, counting every clock cycle to determine the duration of an observation interval is critical however. As the voltage adaptation is done during normal operation, phases with low activity would result in too aggressive voltage scaling. In the extreme case, for an activity rate of zero during an observation interval, no pre-errors occur and the voltage would be lowered independent of the operating condition of the circuit. This will probably lead to timing errors afterward. Besides detecting pre-errors, it is thus also essential to observe the activity of the circuit. Therefore, an additional transition detector to observe the activity rate of the circuit is introduced [43]. It detects transitions during the whole clock cycle, no matter if they are early transitions or pre-errors. As shown in the next chapter, we integrated the capability to detect pre-errors as well as transitions into one single Pre-Error flip-flop. Thus, the Pre-Error flip-flop flags late transitions by the pre-error signal and any switching activity by the transition signal. Adding the transition detector to the pre-error detector is a key innovation and makes the voltage scheme viable for use during normal operation.

4.2.3 AVS Control Unit

The AVS control unit is an essential building block of the voltage control loop, shown in Fig. 4.2, and was developed from scratch during this thesis. It processes the timing information obtained from the in-situ delay monitors. The pre-error as well as the transition signals of the observed outputs are combined by two OR-trees and then fed into the AVS control unit, illustrated in Fig. 4.6. Further inputs are the thresholds $n_{limit\downarrow}$ and $n_{limit\uparrow}$ for the pre-error count as well as the clock signal.

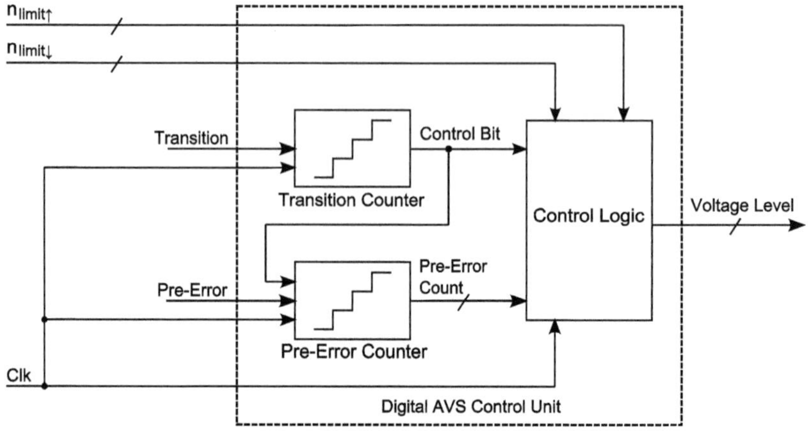

Fig. 4.6 The digital AVS control unit counts the pre-errors during an observation interval and decides whether to decrease, increase or maintain the voltage. The communication to the voltage regulator is done via the binary coded Voltage Level control word

The AVS control unit is purely digital and consists of a pre-error counter, a transition counter and a control logic that triggers the on-chip voltage regulator to increase, decrease or maintain the voltage after every observation interval. The communication to the voltage regulator is performed via a binary control word, called Voltage Level. The digital AVS control unit was composed in VHDL and synthesized in a way to safely operate for the defined voltage range of the AVS scheme from 1.2 V down to 0.8 V. The underlying control algorithm is depicted as flowchart in Fig. 4.7. To reduce the power overhead introduced by the AVS control unit, its design was optimized for power consumption. The power saving potential of the Pre-Error AVS, including all overheads will be discussed thoroughly in Chap. 7.

4.2.4 Voltage Regulator

In SoCs, the power management is typically also integrated on-chip and supplies all building blocks with one or several operating voltages. For this purpose, either switching or linear regulators (or a combination of both) are integrated on the die. Especially in battery powered devices, often linear low-dropout regulators (LDO) are used to provide a stable supply voltage for the digital cores [46, 47]. While batteries discharge over time and hence their output voltages diminish, LDOs provide a constant voltage.

Adaptive voltage scaling naturally requires a voltage regulator that is capable of changing its output voltage with a specified step size ΔV_{DD}. The state-of-the-art LDO architecture [46] can be easily modified to make it practicable for AVS systems. Instead of the 1:1 voltage divider in the feedback loop, which halves the voltage V_{DD}, we propose to implement a resistor ladder. Therefore, the upper resistor of the voltage divider has to be partitioned into multiple resistors, each equipped

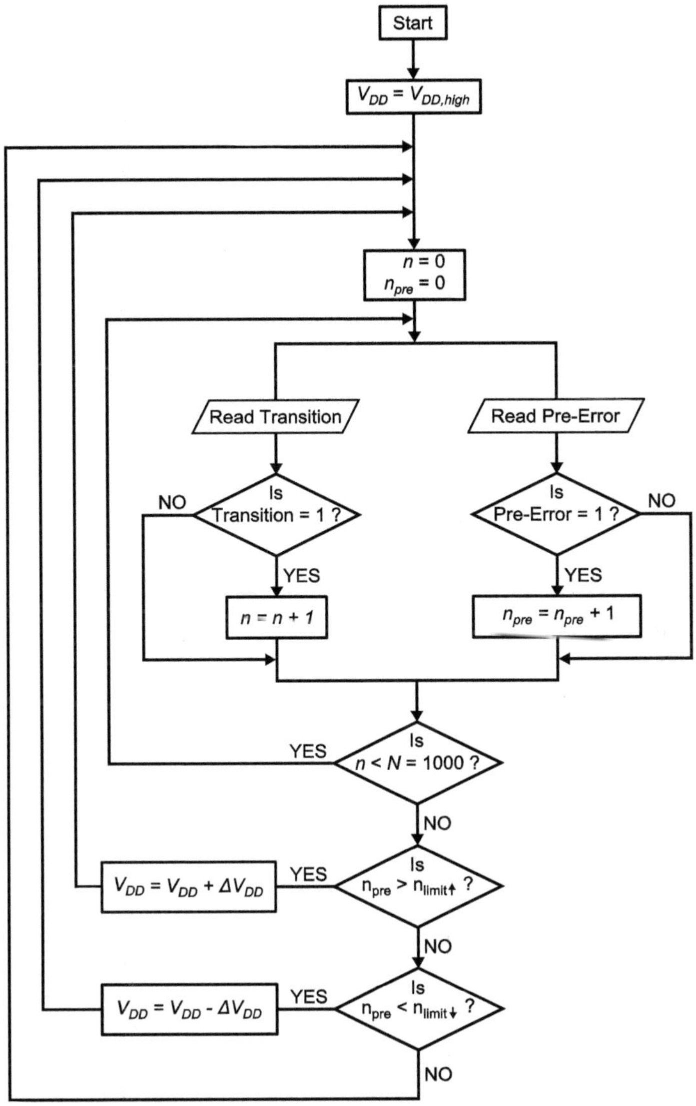

Fig. 4.7 Flowchart of the Pre-Error AVS control algorithm

with a switch, see Fig. 4.8. As the total resistance of the resistor ladder remains unchanged compared to the 1:1 voltage divider, only the switches and the additional wiring lead to some area overhead.

In the shown example, the reference voltage of the regulator is $V_{ref} = 0.6$ V and it is connected to the negative input of the error amplifier. Due to the negative feedback, the positive and negative input of the error amplifier are almost equal in the steady-state. Thus, the node V_+ in the resistor ladder is held constant at 0.6 V.

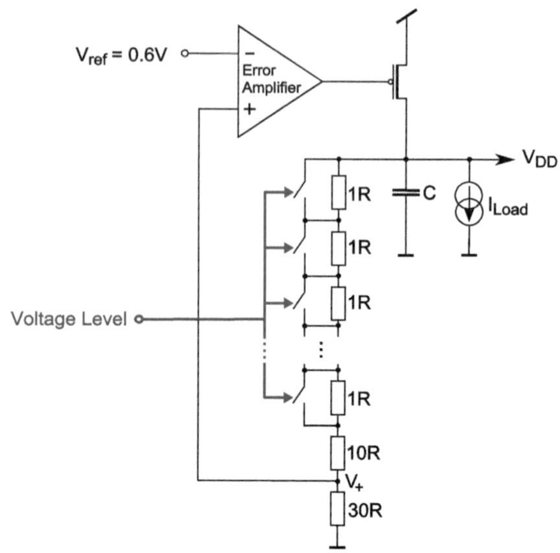

Fig. 4.8 Tunable low dropout regulator (LDO) with resistor ladder at V_{DD}. Each switch is controlled individually by the Voltage Level control word, coming from the AVS control unit

If all 20 switches of the resistor ladder are open, the whole ladder has a resistance of 60 R which leads to a voltage value of 1.2 V at V_{DD}.

For the Voltage Level control word a thermometer code is used and each switch is controlled individually. In the example, the control word has 20 digits b_i ($i = 0 \ldots 19$). For $b_i = 1$, the corresponding switch is open, for $b_i = 0$, it is conducting. Thus, the voltage level at V_{DD} can be calculated by

$$V_{DD} = 0.6\ \mathrm{V} \cdot \frac{40\ \mathrm{R} + \sum_{i=0}^{19} b_i \mathrm{R}}{30\ \mathrm{R}} \tag{4.1}$$

With all switches closed, a voltage of $V_{DD} = 0.8$ V is set. Between the minimum voltage of 0.8 V and the maximum of 1.2 V, the voltage can be set in steps of $\Delta V_{DD} = 20$ mV. With different partitioning of the resistor ladder, the voltage range and the step size can be designed as desired.

The switching time from one voltage level to another depends on the speed of the voltage regulator. These dynamics have to be considered when analyzing the AVS system. Therefore, the effect of the regulator's time constant on the stability of the overall AVS control loop is discussed in Sec. 6.3.

Chapter 5
Design of In-situ Delay Monitors

The in-situ delay monitors used for our AVS approach are conventional flip-flops with additional circuitry to detect pre-errors (late data transitions). Therefore, we refer to these in-situ delay monitors also as Pre-Error flip-flops. Besides pre-error detection, these flip-flops are also capable of detecting early transitions in order to observe the activity rate of the circuit. Note that early transitions are transitions that occur before the pre-error detection window.

For realizing the pre-error detection window, either the duty-cycle of the clock signal or a delay element can be used. In this chapter three different designs to implement the Pre-Error flip-flop are discussed in detail. In the first design, called Crystal-ball flip-flop [41], a delay element comprising an inverter chain defines the pre-error detection window. In the other two designs, the duty-cycle of the clock signal is exploited as the detection window. The idea to define the pre-error detection window by using the duty-cycle was created during this thesis. In these approaches the Pre-Error flip-flop is implemented in dynamic [43] and static design style [44] and named as Dynamic Pre-Error and Static Pre-Error flip-flop, respectively. All designs were implemented in an industrial 65 nm low power CMOS technology. In order to evaluate the accuracy and robustness of the detection window length, which is a key design parameter for the Pre-Error AVS system, the Pre-Error flip-flops were simulated thoroughly on transistor level [48].

5.1 Delay-Element Based Pre-Error Flip-Flop

In the delay-element based Pre-Error design a shadow flip-flop with delayed data input is added in parallel to the regular flip-flop [41]. Figure 5.1 shows the schematic of this so-called Crystal-ball flip-flop. The delay element comprising an inverter chain specifies the length of the detection window. When a data transition occurs closer to the clock rising edge than the propagation time T_{pre} through the delay chain, the shadow flip-flop will miss to latch the correct input data and a pre-error pulse is generated.

M. Wirnshofer, *Variation-Aware Adaptive Voltage Scaling for Digital CMOS Circuits*, Springer Series in Advanced Microelectronics 41, DOI 10.1007/978-94-007-6196-4_5,

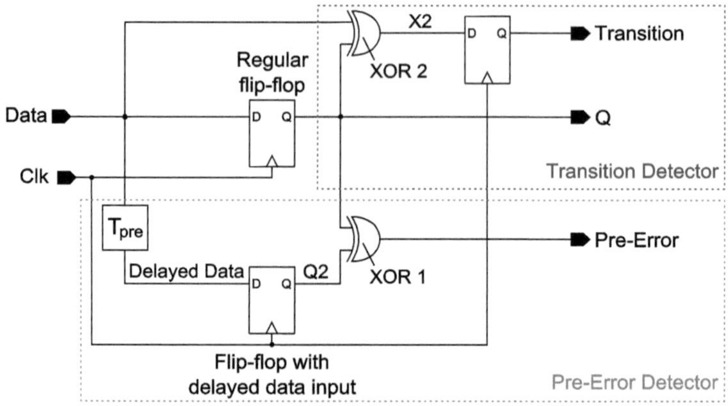

Fig. 5.1 Schematic of the delay element based Pre-Error flip-flop, which is also known under the name Crystal-ball flip-flop

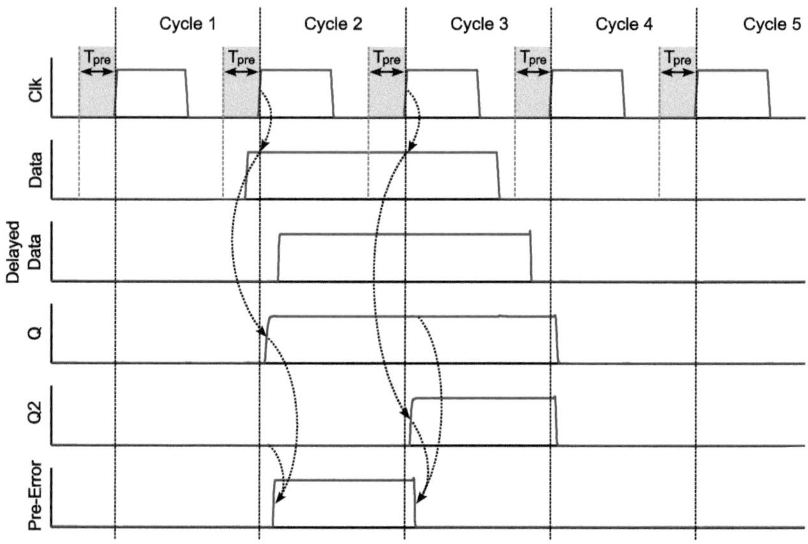

Fig. 5.2 Simulated timing diagram of the delay element based Pre-Error flip-flop, showing the pre-error detection. Transitions inside the pre-error detection window are detected as pre-errors (Cycle 1), whereas early transitions remain undetected (Cycle 3)

For the Pre-Error Detector in the lower part of the schematic, the simulated timing diagram is depicted in Fig. 5.2. In Cycle 1, the data transition occurs during the pre-error detection window but before the setup time of the regular flip-flop. Thus, the regular flip-flop latches the correct value (a logic one), whereas the shadow flip-flop latches a logic zero, as the "Delayed Data" changes only after the rising edge of the clock. In consequence, the output of the regular flip-flop Q and the output of the shadow flip-flop Q2 differ and the XOR1-gate signals a pre-error. The pre-error

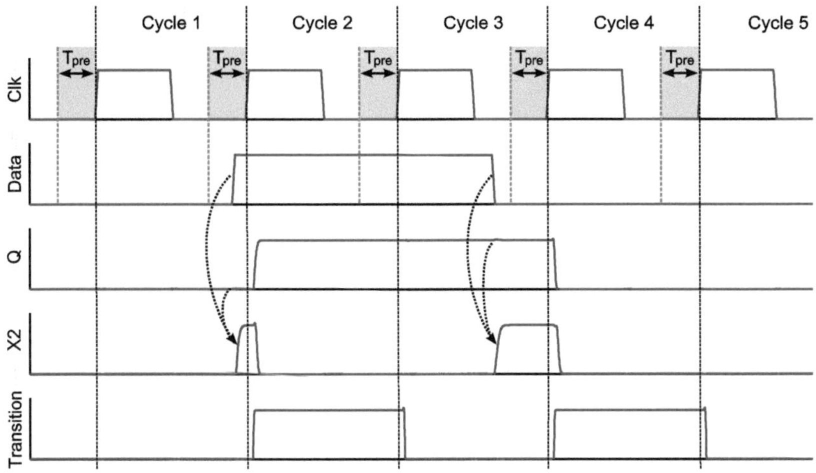

Fig. 5.3 Simulated timing diagram of the delay element based Pre-Error flip-flop showing the transition detection. Every transition is detected, no matter if it is a late transition (Cycle 1) or early transition (Cycle 3). With the Transition detector the activity of the circuit is observed

signal is valid for one clock cycle and is captured with the next rising clock edge by the AVS control unit. For transitions before the detection window, see Cycle 3, regular and shadow flip-flop capture both the correct data and consequently no pre-error shows up.

For the Transition Detector in the upper part of the schematic, the corresponding timing diagram is depicted in Fig. 5.3. The Transition Detector monitors the activity of the circuit by detecting late as well as early transitions. Therefore, the XOR2-gate compares the captured output value Q with the actual input data of the regular flip-flop. In Cycle 1, the output Q is low from the previous clock period and the input data rises during the pre-error detection window. With unequal inputs, the output X2 of the XOR-gate rises and remains at logic one until the regular flip-flop latches its input with the next rising clock edge. This leads to a positive pulse at node X2, which signals the data transition. As mentioned before, all data transitions are detected, independent from the time point of occurrence during the clock cycle. Thus, also transitions before the detection window, see Cycle 3, are monitored and result in a positive pulse at node X2.

As can be seen in Fig. 5.3, the pulse-width at node X2 depends on the time point of the data transition. As the transition pulse is latched by the subsequent flip-flop, it is however only important that X2 is stable during the setup time of this flip-flop. With latching the XOR-output, the transition signal is valid for the following clock cycle and is processed by the AVS control unit with the next rising clock edge.

Note that for the Transition detector, the input data have to propagate through the XOR-gate before arriving at the subsequent flip-flop. Thus, very late data transitions, which arrive just before the setup time of the regular flip-flop, are not captured by the transition detector. Such transitions are, however, extremely seldom and hardly affect the observed activity rate of the circuit. When equipping multiple

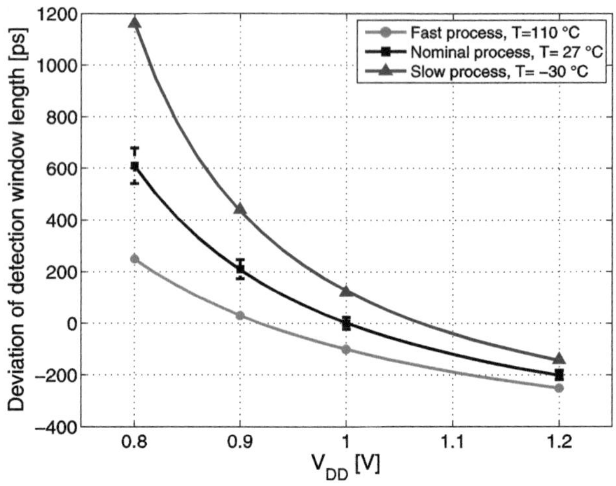

Fig. 5.4 Deviation of the detection window length for the Crystal-ball flip-flop. The graph shows the deviation over V_{DD} for the corner cases. In the nominal corner the detection window length at $V_{DD} = 1.0$ V is $T_{pre} = 600$ ps. The error-bars display the 3σ uncertainty due to local process variations for the nominal corner

outputs with Pre-Error flip-flops, which is the common case, undetected activity of the circuit due to missed transitions gets even less likely.

As the length of detection window is crucial for the voltage scaling, we analyze its accuracy and robustness in the following. The delay element as well as the shadow flip-flop are prone to variations and so is the detection window length T_{pre}. Figure 5.4 illustrates the deviation of T_{pre} over V_{DD} for the fast, nominal and slow corner. The shown curves were obtained by transistor level simulations for a Crystal-ball flip-flop with a nominal detection window length of $T_{pre} = 600$ ps at $V_{DD} = 1.0$ V. To achieve this nominal detection window length, a delay element with 44 inverters is used. The included error bars for the nominal corner show the 3σ uncertainty due to local process variations. These values were derived from Monte-Carlo SPICE simulations with statistic parameters based on measurements.

The detection window has a large shift in the corner cases and a strong dependence on the supply voltage. Our research showed, that for global changes of process, voltage or temperature, the effect of setup time variations of the shadow flip-flop is very small compared to the variations of the delay element.

The 3σ uncertainty due to local variations is 69 ps at $V_{DD} = 0.8$ V and gradually decreases to 18 ps at $V_{DD} = 1.2$ V. Here, the contribution of setup time variations of the shadow flip-flop and variations of the delay element are similar. Note that the quantity of global and local deviations of the delay element depend on the nominal length of the detection window. For longer detection windows, the inverter delay chain is longer and hence the deviations are larger and vice versa. The shadow flip-flop's setup time variations are independent of the detection window length.

5.2 Duty-Cycle Based Pre-Error Flip-Flops

Instead of using a delay element for the pre-error detection window, also the duty-cycle can be exploited to define the detection window length. Exploiting the duty-

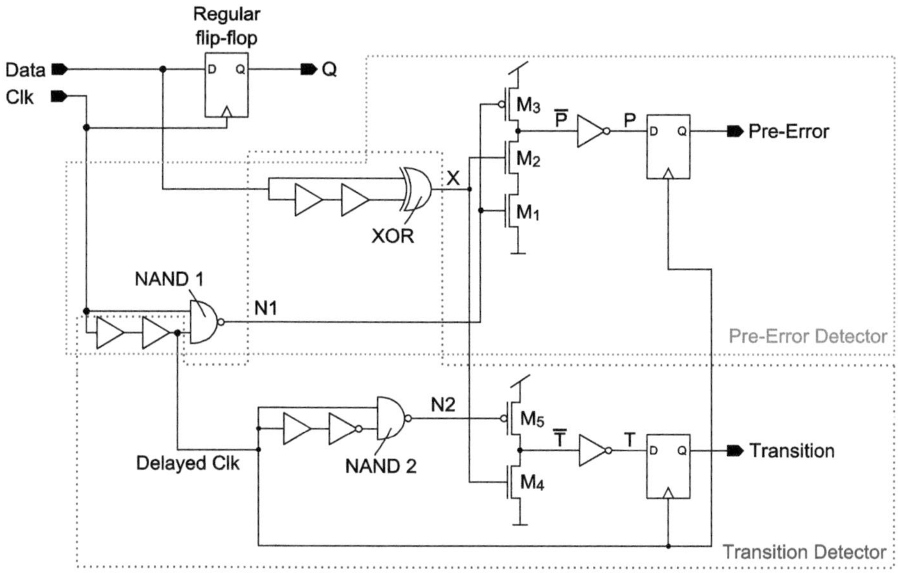

Fig. 5.5 Schematic of the duty-cycle based Pre-Error flip-flop designed in dynamic logic style (Dynamic Pre-Error flip-flop)

cycle means that the detection window starts with the falling clock edge. By changing the duty cycle, e.g. by a multiphase clock [49, 50] or a pulse-width-control loop (PWCL) [38, 50], T_{pre} can be adjusted. This way the threshold that separates early and late transitions can be variably set. Arbitrarily changing the duty-cycle is, however, only possible if the falling clock edge is not used to trigger any logic events. The following two sections show a dynamic as well as static implementation for the novel duty-cycle based Pre-Error flip-flops.

5.2.1 Dynamic Implementation

The schematic of the duty-cycle based Pre-Error flip-flop designed in dynamic logic is shown in Fig. 5.5. For the sake of brevity it is simply called Dynamic Pre-Error flip-flop in the following.

The XOR-gate in the upper part of the schematic is a central part of the circuit. For every data transition, the first input of the XOR-gate changes its value immediately, whereas the second input will change its value only with a short delay. This leads to a pulse at the output node X of the XOR-gate, see Cycle 1 and Cycle 3 in Fig. 5.6. The duration of these pulses is dependent on the generated data delay at the second input of the XOR-gate.

The NAND1-gate generates an inverted clock with prolonged detection window at its output N1. This inverted clock has a longer high phase, hence all data

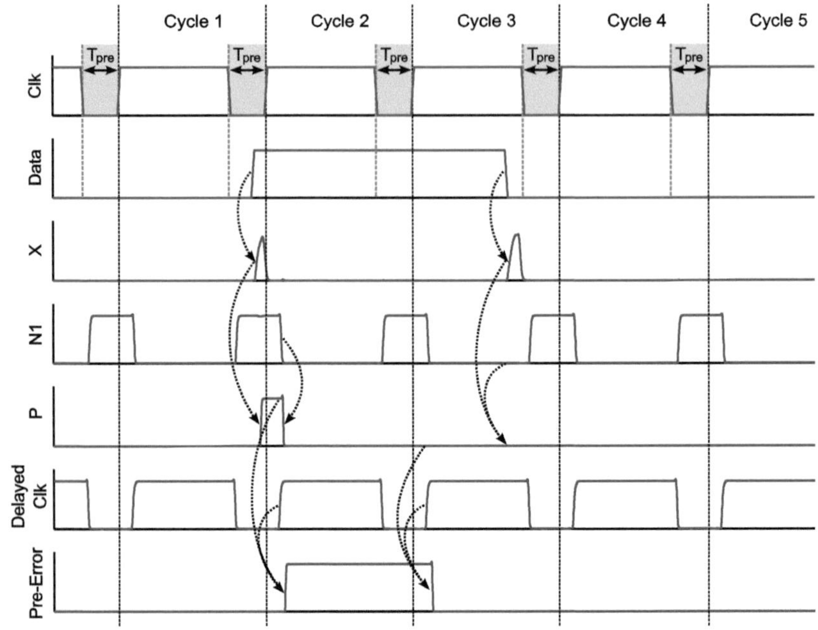

Fig. 5.6 Simulated timing diagram of the Dynamic Pre-Error flip-flop, showing the pre-error detection. Data transitions lead to a pulse at node X and are detected as pre-error if they occur during the detection window (Cycle 1). Early transitions lead to a pulse before the detection window and remain undetected (Cycle 3)

transitions inside the pre-error detection window are detected regardless of process and voltage variations. With transistor M_1 conducting during the high phase of N1 (detection window), the XOR-pulse induces a conducting path from node $\overline{P}$ to ground in case of a late transition. Node $\overline{P}$ is pulled down and the inverted node P is latched by the delayed clock signal to have a pre-error signal, which is valid for the complete next clock cycle. The latched pre-error is then fed into the AVS control unit.

During the high-clock phase, node N1 goes low and transistor M_1 is turned off. At the same time transistor M_3 is turned on and node $\overline{P}$ is reset to V_{DD}. With transistor M_1 in off-state, transitions during this time span are not detected as pre-errors. As mentioned before, the detection window length, denoted by T_{pre}, is defined by the falling clock edge and can thus be adjusted by changing the duty cycle.

The transition detector in the lower part of the schematic is also implemented in dynamic design style. Its functionality is illustrated by the corresponding timing diagram in Fig. 5.7.

The transition detector signals every data transition and therefore reuses the XOR-pulse at node X. The XOR-pulse turns on transistor M_4 and induces a conducting path from node $\overline{T}$ to ground. Node $\overline{T}$ is pulled down and the inverted node T is latched with the delayed clock signal to have a transition signal, which is valid for the complete next clock cycle. The latched transition signal is then fed into the AVS

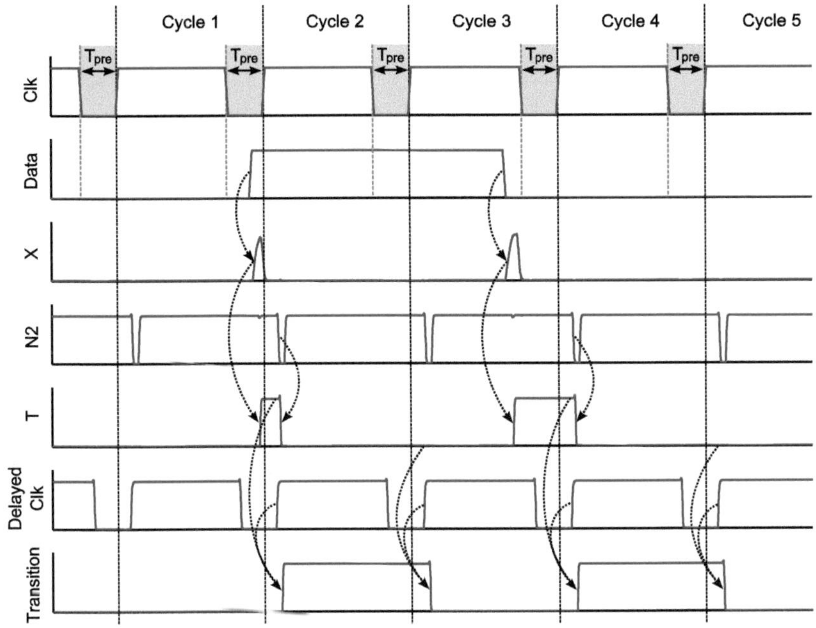

Fig. 5.7 Simulated timing diagram of the Dynamic Pre-Error flip-flop showing the transition detection. Data transitions lead to a pulse at node X and are detected, no matter if it is a late (Cycle 1) or early transition (Cycle 3). With the Transition detector the activity of the circuit is observed

control unit. Shortly after node T is latched, the NAND2-gate generates a short low pulse at node N2 for resetting node $\overline{\mathrm{T}}$ to V_{DD} and node T to ground, respectively. This reset pulse is delayed to avoid a situation where transistor M_4 and M_5 are both conducting and a cross-current would flow.

When analyzing the robustness of the detection window for the Dynamic Pre-Error flip-flop, not only the variations of the Pre-Error detector itself have to be considered, but also variations of the clock-tree affecting the clock signal (duty-cycle). For our evaluation, we use a conventional H-tree with three buffer stages, depicted in Fig. 5.8, as the H-tree is the basic clock topology for many clock distribution systems [51].

Figure 5.9 shows the sensitivity of the detection window length under global and local variations. The results were obtained by Corner and Monte-Carlo simulations on transistor level. For the deviation of the detection window, variations of the Dynamic Pre-Error flip-flop itself and the clock-tree were considered.

For the Dynamic Pre-Error flip-flop, the deviations of T_{pre} for the corner cases and over V_{DD} are small compared to the Crystal-ball flip-flop. This means, the dynamic Pre-Error detector in combination with the clock-tree is more robust against global variations than the delay element of the Crystal-ball flip-flop.

The 3σ uncertainty due to local variations is very similar to the Crystal-ball flip-flop and reads as 91 ps at $V_{DD} = 0.8$ V and 18 ps at $V_{DD} = 1.2$ V. Regarding the local variations, the Pre-Error detector accounts for approximately two third and

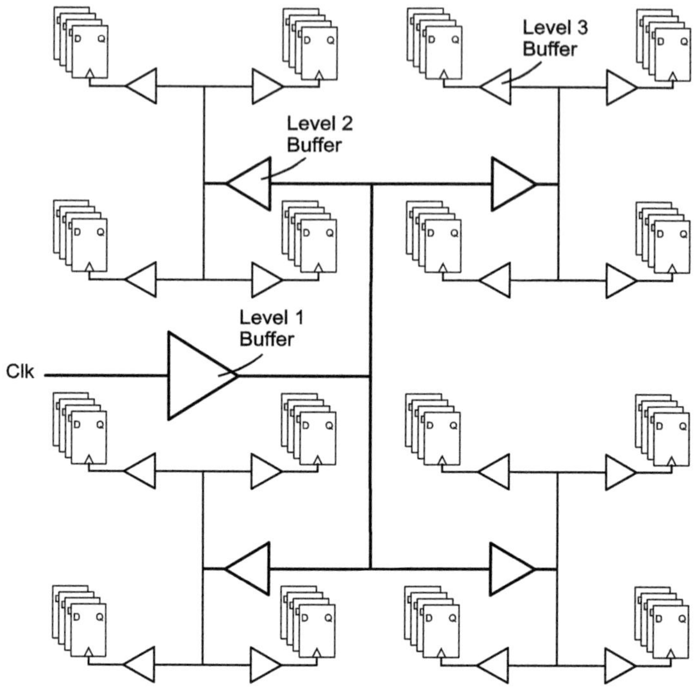

Fig. 5.8 H-tree for clock distribution with three levels of buffer stages. The H-tree is the most common clock tree topology and was used to analyze the effect of PVT variations on the duty-cycle

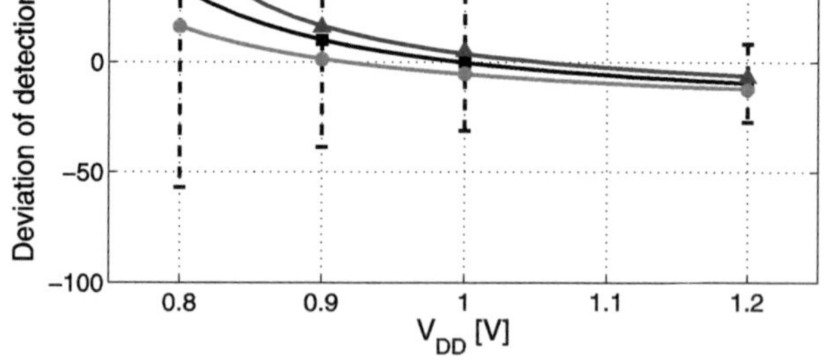

Fig. 5.9 Deviation of the detection window length for the Dynamic Pre-Error flip-flop. The graph shows the deviation over V_{DD} for the corner cases. The error-bars display the 3σ uncertainty due to local process variations for the nominal corner. For the Dynamic Pre-Error flip-flop, the global and local deviations are independent of the detection window length

the clock-tree for one third of the uncertainty. Note that the global as well as local deviations are independent of the detection window length. Thus, the longer the detection window length, the smaller are the relative deviations.

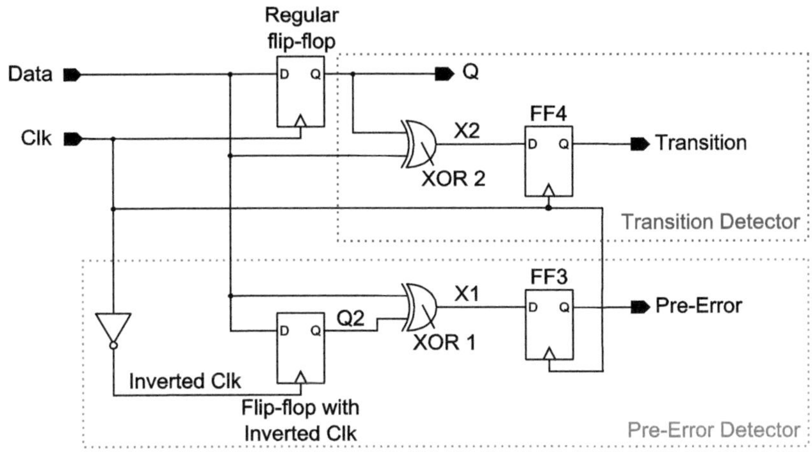

Fig. 5.10 Schematic of the Static Pre-Error flip-flop, consisting of Pre-Error and Transition detector. It is designed with standard cells only and can be implemented very quickly in any new technology

5.2.2 Static Implementation

Dynamic circuits are typically used when high performance is required. They are, however, not ratio-less and therefore have to be designed very carefully, especially in nanometer technologies with increased process uncertainties. To reduce design time and avoid the risk of costly redesigns, dynamic logic is more and more replaced by static CMOS in recent digital systems [52].

Therefore, we also designed a static version of the duty-cycle based Pre-Error flip-flop as an alternative to the dynamic one. The schematic of the Static Pre-Error flip-flop can be seen in Fig. 5.10. It contains only standard cell elements and can hence be implemented very quickly in any new technology.

Besides the regular flip-flop, the Pre-Error Detector (lower part of Fig. 5.10) latches the incoming data with the falling clock-edge. The sampled value Q2 is compared with the present input data by the XOR1-gate. If a data transition occurs during the low clock phase, signals Q2 and Data differ and the XOR1 output X1 goes high, see Fig. 5.11 (Cycle 1). The output X1 is then latched with the next rising clock edge by the subsequent flip-flop (FF3) to have a stable pre-error signal during the next clock cycle. If the data gets valid already during the high clock phase, no pre-error is flagged, see Fig. 5.11 (Cycle 3). Note that transitions occurring one XOR-delay before the setup time of FF3 are not detected. This, however, has typically a marginal effect, since pre-errors rather occur at the beginning of the detection window than at the end, see Fig. 4.4 (on page 26). For very short detection windows, where this limitation might be an issue, the clock signal of FF3 can be easily shifted by a delay element. Then, all transitions till the end of the clock cycle are covered.

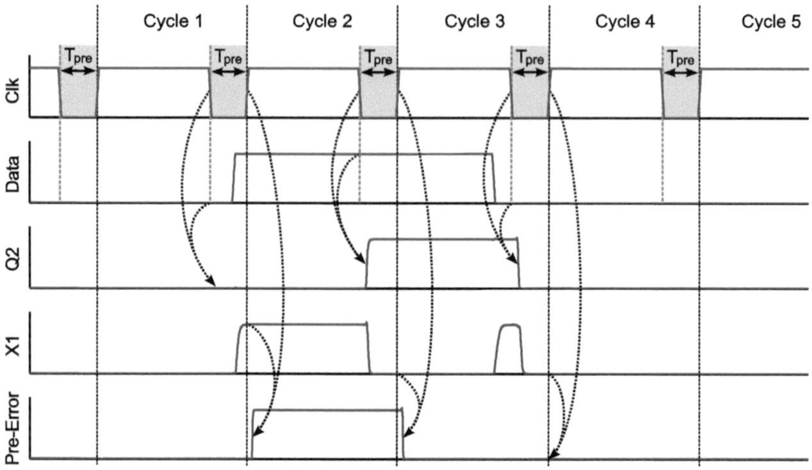

Fig. 5.11 Simulated timing diagram of the Static Pre-Error flip-flop, showing the Pre-Error detection. The late transition during the detection window in Cycle 1 leads to a pre-error. The early transition before the detection window in Cycle 3 is not detected

The Transition Detector in the upper part of Fig. 5.10 signals transitions from one rising clock edge to the next. It is implemented in the same way as for the Crystal-ball flip-flop in Fig. 5.1. If the data input changes, the sampled data Q and the present input D differ. The XOR2-gate detects this difference and X2 goes high, see Fig. 5.3 (Cycles 1 and 3). The X2 signal is captured by the subsequent flip-flop (FF4) to have a stable transition signal during the next clock cycle. In Cycles 2 and 4 no transition occurs and the transition signal goes down again. Similar to the Pre-Error detector, transitions occurring one XOR-delay before the setup time of FF4 are not detected. The transition detection is, however, even less affected by this limitation than the Pre-Error detection. In the case that the clock signal is already delayed for FF3, this shifted clock can also be used for FF4 to detect all transitions.

When analyzing the robustness of the Static Pre-Error flip-flop, the same clock tree was assumed as for the Dynamic Pre-Error flip-flop. Figure 5.12 illustrates the deviations of the detection window when considering variations of the Static Pre-Error flip-flop itself and the clock-tree.

For the Static Pre-Error detector, global PVT variations hardly affect the detection window length as the duty-cycle and the setup time stay almost constant. Hence, its deviation stays below 10 ps for all corners, providing very reliable Pre-Error detection.

The 3σ uncertainty due to local variations reads as 49 ps at $V_{DD} = 0.8$ V and 15 ps at $V_{DD} = 1.2$ V. Thus, the Static Pre-Error flip-flop is less sensitive to local variations than the Dynamic Pre-Error flip-flop. For the Static Pre-Error flip-flop, deviations of the detection window over V_{DD} or due to process/temperature variations are independent of the nominal detection window length, just as for the Dynamic Pre-Error flip-flop.

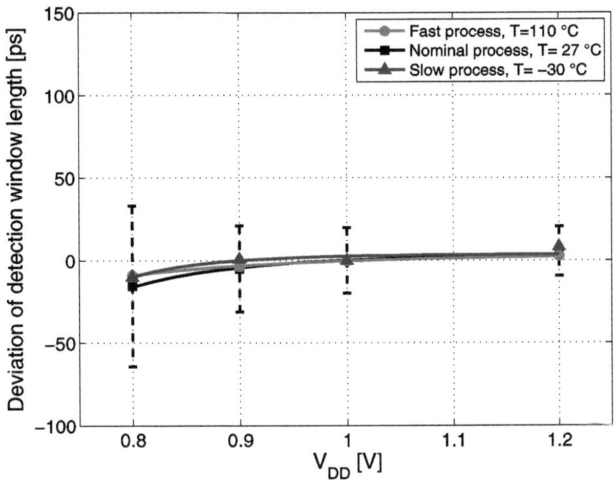

Fig. 5.12 Deviation of the detection window length for the Static Pre-Error flip-flop. The graph shows the deviation over V_{DD} for the corner cases. The error-bars display the 3σ uncertainty due to local process variations for the nominal corner. The global and local deviations are independent of the detection window length

5.3 Comparison of In-situ Delay Monitors

The basic characteristics of the three designed Pre-Error Flips-Flops are summarized in Table 5.1. The Crystal-ball flip-flop is based on a delay element, whereas the Dynamic and the Static Pre-Error flip-flop exploit the duty-cycle of the clock to define the detection window length T_{pre}. Implementing the Static Pre-Error flip-flop requires the lowest design effort, as it only consists of static standard cells. Also the Crystal-ball flip-flop is mainly built out of standard cells, only the delay element has to be customized for the desired detection window length. The dynamic implementation of the in-situ delay monitor is the most complex design. The pull-up and pull-down transistors have to be sized very carefully to achieve a robust design for all corners. The initial reason to still implement a dynamic pre-error detector was its capability to detect also glitches inside the detection window. The Static Pre-Error flip-flop as well as the Crystal-ball flip-flop miss these glitches, resulting in undetected pre-errors. This, however, has a marginal effect, as critical transitions rather occur at the beginning of the detection window and thus glitches inside the detection window are extremely unlikely. Extensive transistor level simulations showed that the number of undetected pre-errors (glitches inside the detection window) accounts typically for only 0.1 % of all pre-errors.

Table 5.1 also shows the power overhead introduced by the three different types of in-situ delay monitors. The power consumptions of the Pre-Error flip-flops were simulated with SPICE to obtain accurate and reliable data. Compared to the regular flip-flop, the Static Pre-Error flip-flop consumes 230 % of additional power for pre-error and transition detection. Regarding the Crystal-ball flip-flop, its power dissipation mainly results from the delay element. For a nominal detection window length of $T_{pre} = 600$ ps, which is typical for a clock period of $T_{Clk} = 2$ ns, the Crystal-ball flip-flop burns 580 % of extra power. Note that for higher operating frequencies, the clock period as well as the detection window get shorter and the power dissipation decreases. Lower operating frequencies consequently result in an increased power

Table 5.1 Summarized key metrics for the different Pre-Error flip-flops

	Crystal-ball	Dynamic	Static
Detection window	delay-element based	duty-cycle based	duty-cycle based
Design style	static[a]	dynamic	static[b]
Design effort	low effort	large effort	very low effort
Extra power compared to conventional flip-flop	580 %[c]	570 %	230 %
Extra area compared to conventional flip-flop	700 %[c]	480 %	270 %
Deviation of T_{pre} under local variations	$3\sigma_{0.8\text{ V}} = 69$ ps[c] $3\sigma_{1.2\text{ V}} = 18$ ps[c]	$3\sigma_{0.8\text{ V}} = 91$ ps $3\sigma_{1.2\text{ V}} = 18$ ps	$3\sigma_{0.8\text{ V}} = 49$ ps $3\sigma_{1.2\text{ V}} = 15$ ps
Shift of T_{pre} at 1.0 V under global variations	slow case: +120 ps[c] fast case: −100 ps[c]	slow case: +5 ps fast case: −5 ps	slow case: +2 ps fast case: −1 ps

[a]With standard cells and delay element
[b]With standard cells only
[c]The values for the Crystal-ball flip-flop are obtained for a nominal detection window length of $T_{pre} = 600$ ps. The power consumption as well as the variations of T_{pre} increase for longer detection windows and vice versa

overhead for the Crystal-ball flip-flop. The additional power for the Dynamic Pre-Error flip-flop reads as 570 % and is—just like for the Static one—independent of the detection window length. Regarding the extra area, again the Crystal-ball flip-flop requires the most overhead (700 %), followed by Dynamic (480 %) and Static Pre-Error flip-flop (270 %).

The effect of local variations on the Pre-Error flip-flops was analyzed by Monte-Carlo SPICE simulations for an industrial 65 nm low-power technology. Our research showed that the resulting distributions of the detection window length are normally distributed for all three Pre-Error flip-flops. Thus, the standard deviation σ of the their distributions can be used as metric for their sensitivity to local variations. The three times standard deviations (3σ) are summarized in Table 5.1. The low 3σ-values show that all three designs are very robust against local variations. The most robust design is the Static Pre-Error flip-flop, for which the three times standard deviation at $V_{DD} = 0.8$ V reads as $3\sigma_{0.8\text{ V}} = 49$ ps. For the Crystal-ball flip-flop and the Dynamic Pre-Error flip-flop the three sigma values are $3\sigma_{0.8\text{ V}} = 69$ ps and $3\sigma_{0.8\text{ V}} = 91$ ps, respectively. The deviations of the Crystal-ball flip-flop are a combination of setup time and delay-element variations and were obtained for a nominal detection window length of $T_{pre} = 600$ ps. A shorter delay element leads to a smaller number of inverters in the delay chain and hence to reduced detection window variations and vice versa. For the Static and Dynamic Pre-Error flip-flop the variations only depend on the clock tree and are not influenced by the nominal detection window length.

Under global variations, the detection window of the Static and the Dynamic Pre-Error flip-flop are hardly affected. In contrast, the detection window of the Crystal-ball flip-flop is shifted by 240 ps from slow to fast corner for a nominal length of $T_{pre} = 600$ ps ($V_{DD} = 1.0$ V). Again, the shift of the Crystal-ball flip-flop is dependent on the nominal detection window length.

The impact of detection window variations on the voltage adaptation is investigated in detail in Chap. 7. Also the power overhead of the Pre-Error flip-flops, which reduces the overall power saving potential of the Pre-Error AVS system, is discussed there. Generally, one can state that for typical operating frequencies the duty-cycle based Pre-Error flip-flops are preferable, as the Crystal-ball flip-flop suffers from its long delay element chain. A long delay chain is power hungry and the detection window defined by its propagation time is sensitive to variations.

Chapter 6
Modeling the AVS Control Loop

As stated before, adapting the supply voltage by using in-situ delay monitors forms a closed-loop control system. The last chapter focused on the in-situ delay monitors (Pre-Error flip-flops) acting as sensors of this system. The following chapter will now deal with the entire control loop. First, we show how the whole system can be analyzed accurately and at the same time efficiently. Subsequently, Sect. 6.2 thoroughly explains the Markov chain, which is used to model the AVS system. Describing the voltage adaptation by a Markov model is a beneficial approach, which was elaborated during this work and published first in [43]. With the Markov model, we are able to evaluate the power saving potential and reliability of the Pre-Error AVS system. In the last section of the chapter, the stability of the AVS control loop is discussed, which has to be ensured for proper system operation.

6.1 Simulation Methodology

Static Timing Analysis (STA) is the method of choice to estimate the path delays of digital semi-custom circuits during synthesis (design stage) [53, 54]. Moreover, STA is also applied for the timing sign-off, i.e. the verification whether all timing requirements are fulfilled in the final design.

In the STA concept, all standard cells are first characterized by transistor level simulations [55]. During characterization, the delay and output slope for each available logic gate is simulated on transistor level for various input slopes and output capacitances. From these simulations, look-up tables for each cell are created, containing the resulting output slopes and propagation delays for different combinations of input slope and load capacitance [54]. The STA uses these look-up tables to model the dynamic behavior of the logic gates inside a digital circuit. This delay model of the STA results in a tremendous reduction in simulation time compared to transistor level simulations, but yields a significant loss in accuracy at the same time. Due to the imprecision of the STA, its timing approximations have to be rather conservative to avoid over-optimistic delay estimations, which would result in timing failures of the manufactured circuit.

M. Wirnshofer, *Variation-Aware Adaptive Voltage Scaling for Digital CMOS Circuits*, Springer Series in Advanced Microelectronics 41, DOI 10.1007/978-94-007-6196-4_6,

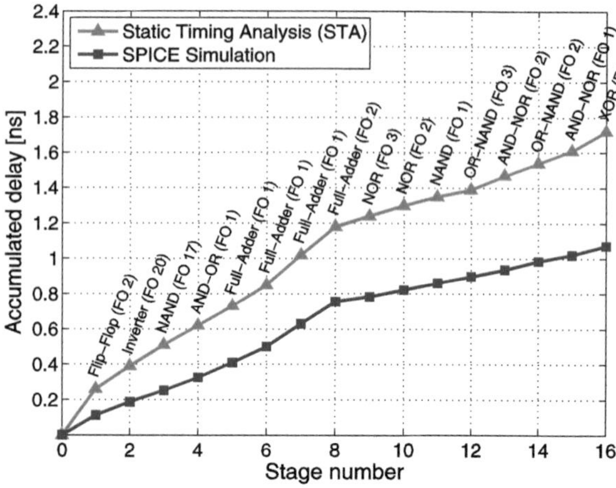

Fig. 6.1 Comparison between Static Timing Analysis (STA) and SPICE simulation for the critical path delay of a synthesized multiplier circuit in 65 nm technology. The accumulated path delay after each logic gate diverges due to the conservative timing estimation of the STA. The fan-out (FO) of each logic gate is given *in brackets*

When comparing the STA with transistor level simulations the difference in accuracy gets clearly visible. For comparison, we use a synthesized 16-bit multiplier circuit in a 65 nm technology and evaluate the most critical path, identified by the STA. We check the path delay estimated by the STA against the SPICE simulated delay of the same path. To have a fair comparison, this is done under identical operating conditions of the circuit (supply voltage $V_{DD} = 1.18$ V, temperature $T = 110\,^{\circ}$C, slow process corner).

The diverging results of STA and SPICE simulation are presented in Fig. 6.1. The y-axis shows the accumulated delay after each logic gate. For the whole path consisting of 16 logic gates (including the flip-flop at the beginning), the STA estimates a delay of 1.72 ns. Thus, it overestimates the actual delay of 1.07 ns, obtained by the SPICE simulation, by 60 %.

We also performed SPICE simulations for 100.000 randomly generated input patterns applied to the multiplier circuit. When analyzing the resulting delays, we found a maximum path delay of 1.14 ns on a path different from the one found by the STA. Thus, we can state, that the STA did not find the real most critical path, but by its cautious timing estimation, the worst-case delay is still by far covered. This pessimism is acceptable for synthesis and timing sign-off, but for evaluating the Pre-Error AVS system, the timing estimations of the STA are not accurate enough. Instead, transistor level simulations are necessary to obtain meaningful results.

Some challenges arise, however, when simulating the Pre-Error AVS system on transistor level. Remember that the AVS control is dependent on the statistics of the output delays and the supply settles to a voltage level only with a certain probability. To determine these probabilities accurately, very long simulation times of the complete system are necessary. This is computationally very intensive for large digital blocks, leading to impracticable long computation run times.

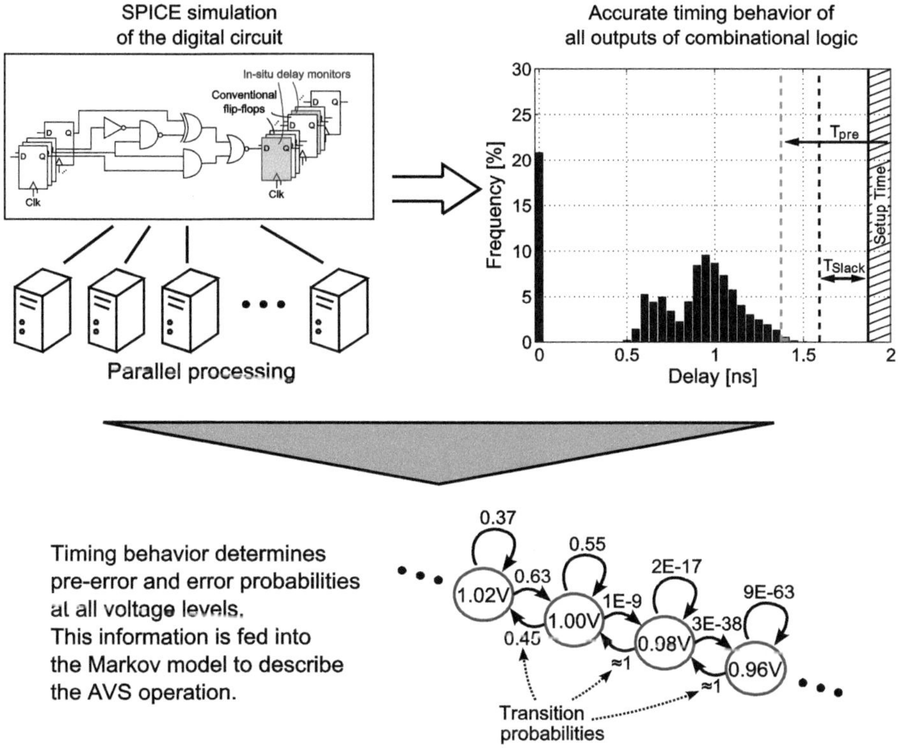

Fig. 6.2 Simulation methodology to model and analyze the AVS control loop: First, SPICE simulations are performed to obtain the accurate timing behavior of the circuit. These simulations are processed in parallel to speed up computation times. The simulation results are used to determine the pre-error and error rates at all voltage levels. The resulting pre-error and error rates are fed into the Markov model to analyze the closed-loop control of the overall AVS system

To solve this problem we developed the simulation methodology shown in Fig. 6.2. We first perform SPICE simulations on the digital block for typical input patterns to get accurate delay statistics for all outputs. In order to obtain the timing behavior over V_{DD}, the same patterns are simulated at different supply voltages. As mentioned before, these simulations are computationally intensive, but by running them on multiple CPUs in parallel, we were able to reduce the computation time tremendously. The extracted delays are then fed into a Markov model that fully describes the closed-loop Pre-Error AVS scheme [43]. The Markov chain is visualized as directed graph in Fig. 6.2. The numbers next to each arrow denote the transition probabilities between the voltage states. More details on the Markov model and how it is adopted to analyze the Pre-Error AVS system are explained in the following section. In contrast to the proposed two-step simulation methodology (first SPICE simulation, than Markov modeling), simulating the whole AVS system directly in SPICE can not be parallelized and is thus computationally infeasible.

6.2 Markov Model of the Pre-Error AVS System

In digital logic, the output delays depend not only on the operating condition of the circuit, but also on the applied input patterns. Thus, for a fixed operating condition, the pre-error count can be seen as a random variable and the voltage is increased or decreased only with a certain probability. To describe the process of voltage adaptation, we developed a discrete time Markov model,[1] described in the following.

6.2.1 Statistics of the Voltage Control

The voltage adaptation based on pre-error detection was already explained in Chap. 4. Having a closer look on the delay distribution for the most critical output of a 16-bit multiplier circuit shall illustrate the working principle once more. The delay distributions were obtained with transistor level simulations, by applying random input patterns to the multiplier circuit. Figure 6.3 shows the resulting distribution at two different supply voltages. For the nominal supply voltage of $V_{DD} = 1.20$ V (Fig. 6.3(a)), the timing is uncritical as the clock period is $T_{Clk} = 2$ ns and the maximum delay is at around $t_{d,max} = 1$ ns. The dashed line denotes the pre-error detection window with a length of $T_{pre} = 600$ ns. It can be seen, that all signal transitions occur before the detection window. Hence, no pre-error occurs and the voltage will be reduced.

By reducing the supply voltage the circuit is slowed down, i.e. the delays increase. This leads to a shifted delay distribution, as shown in Fig. 6.3(b) ($V_{DD} = 0.98$ V). In this constellation the maximum delay increases to around $t_{d,max} = 1.6$ ns and delays longer than $t_d = T_{Clk} - T_{pre} = 1.4$ ns, i.e. the part of the distribution lying right of the dashed line, will result in a pre-error. Remember that the voltage is only reduced for a pre-error count below the threshold $n_{limit\downarrow}$ during an observation interval. Consequently, a voltage reduction gets rather unlikely compared to the situation in Fig. 6.3(a).

The probability for a pre-error count n_{pre} below $n_{limit\downarrow}$ during an observation interval of N clock cycles and consequently the probability $P_{V_{DD}\downarrow}$ that the AVS control unit decides to reduce the supply voltage is given by

$$P[n_{pre} < n_{limit\downarrow}] = P_{V_{DD}\downarrow} = \sum_{n_{pre}=0}^{n_{limit\downarrow}-1} \binom{N}{n_{pre}} \cdot (P_{pre})^{n_{pre}} \cdot (1 - P_{pre})^{N-n_{pre}} \quad (6.1)$$

Each addend in Eq. 6.1 gives the probability of getting exactly n_{pre} pre-errors during an observation interval (see a textbook on statistics like [64]). By summing

[1] The Markov model is named after Andrei Markov, who developed his famous theory about stochastic processes in 1906 [56]. Since than the Markov chain has been applied in many scientific fields, like biology [57, 58], information theory [59, 60], finance [61] and even in music for algorithmic composition [62, 63].

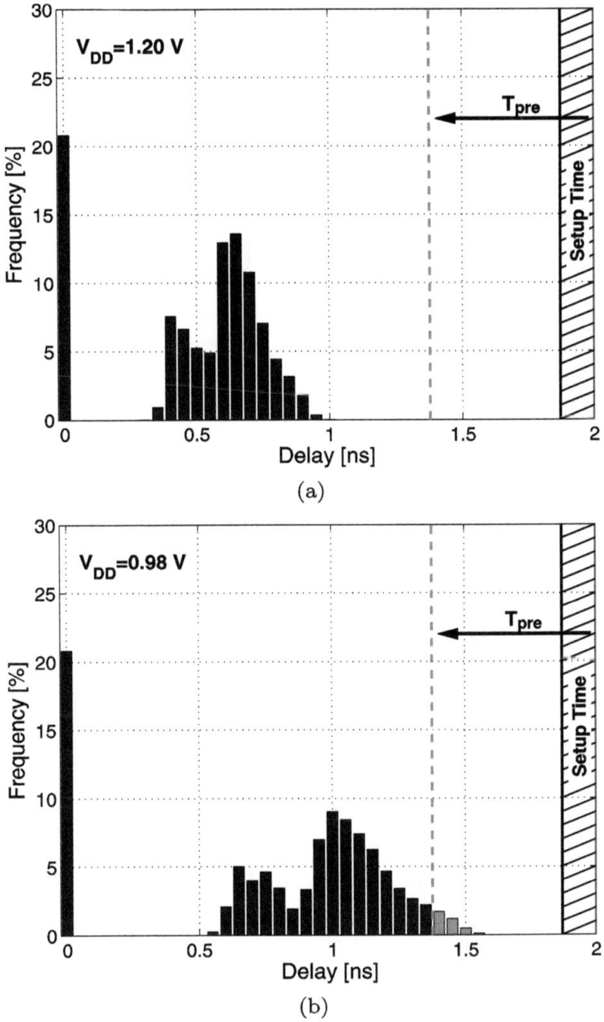

Fig. 6.3 Delay distribution for the most critical output, when applying random patterns to a multiplier circuit. The histograms were obtained by transistor level (SPICE) simulations at a supply voltage of (**a**) $V_{DD} = 1.20$ V and (**b**) $V_{DD} = 0.98$ V (nominal process, $T = 27$ °C)

up all addends from $n_{pre} = 0$ through $n_{pre} = n_{limit\downarrow} - 1$ we obtain the probability $P[n_{pre} < n_{limit\downarrow}]$. For a given delay distribution, this probability is directly dependent on the clock cycles N of the observation interval, see Eq. 6.1. Moreover, it is dependent on the length of the detection window, as T_{pre} directly affects the pre-error rate P_{pre}. Selecting an appropriate number N of clock cycles for the observation interval and choosing the length of the detection window T_{pre} is hence a key task when designing the AVS system.

To avoid the risk of timing errors, the supply voltage should never be lowered to a point where the maximum delay violates the setup time of the flip-flops. On the other hand, unnecessary safety margin should be reduced as far as possible to save power. According to this, if the timing slack is large, a voltage reduction should be likely,

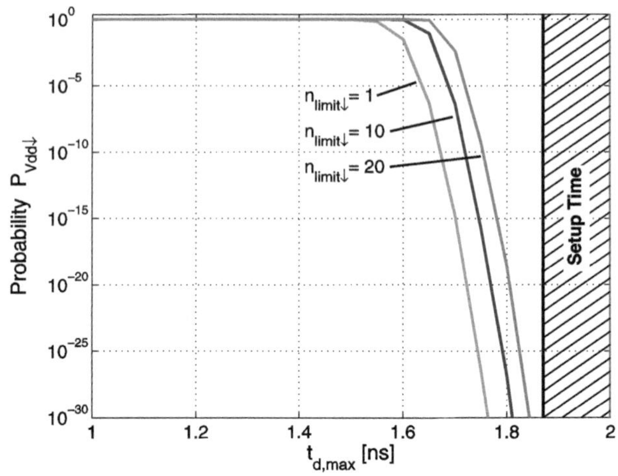

Fig. 6.4 Probability $P_{V_{DD}\downarrow} = P[n_{pre} < n_{limit\downarrow}]$ over the maximum delay $t_{d,max}$ for a detection window length of $T_{pre} = 600$ ps and an observation interval of $N = 1000$ clock cycles

i.e. the probability $P_{V_{DD}\downarrow}$ should be high. This way, unnecessary safety margin is cut and energy consumption is reduced. On the other hand, if the timing slack is low, $P_{V_{DD}\downarrow}$ should also be low to avoid further voltage reductions that provoke timing errors.

We will take the 16-bit multiplier circuit as an example to quantify the probability $P_{V_{DD}\downarrow}$ for reducing the supply voltage for different operating conditions. Therefore, the three most critical outputs of the multiplier are equipped with Pre-Error flip-flops. Furthermore, we choose a detection window length of $T_{pre} = 600$ ps and an observation interval of $N = 1000$ clock cycles. For this detection window length, we determine the pre-error probability P_{pre} for each voltage level by simulating the delays of 100.000 random input patterns. For every voltage level we also know the corresponding maximum output delay. With the extracted values for P_{pre} we can calculate the probability $P_{V_{DD}\downarrow} = P[n_{pre} < n_{limit\downarrow}]$ by Eq. 6.1. Figure 6.4 shows the resulting probabilities in dependence of the maximum output delay $t_{d,max}$ for different threshold limits $n_{limit\downarrow}$. The corresponding timing slack is calculated by $T_{slack} = T_{Clk} - t_{Setup} - t_{d,max}$.

It can be seen that for low maximum delays (large timing slacks), the probability $P_{V_{DD}\downarrow} = P[n_{pre} < n_{limit\downarrow}]$ is almost one. Here, the supply voltage will be reduced very likely. For lower timing slacks, the probability $P_{V_{DD}\downarrow}$ decreases steeply. A further reduction of V_{DD} is hence very unlikely. The three plotted curves show the influence of the threshold value $n_{limit\downarrow}$. For a larger threshold, the probability $P_{V_{DD}\downarrow}$ increases and the voltage is reduced more aggressively.

Reverting to the AVS scheme, we see that the pre-error count is a suitable measure for controlling the supply voltage. If the timing slack is large, due to process or temperature changes, the voltage will be reduced very likely. On the other hand, if the timing gets critical, as a result of variations or previous voltage lowering, a voltage reduction becomes very unlikely.

As the pre-error count is based upon a stochastic process, an overcritical reduction can never be completely ruled out. We can, however, adjust this probability in a

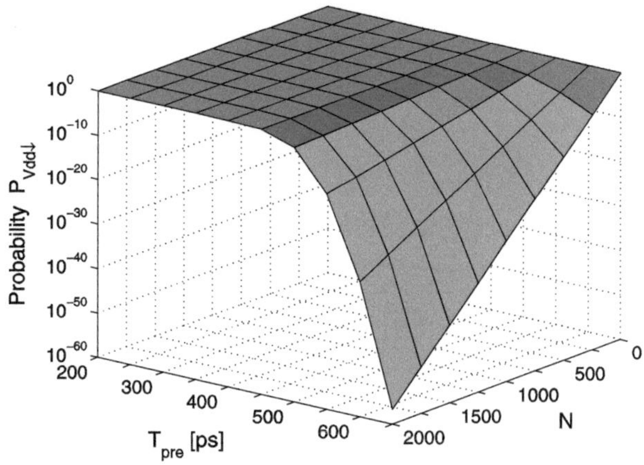

Fig. 6.5 Probability $P_{V_{DD}\downarrow}$ in dependence of the detection window length T_{pre} and the number of clock cycles N for the observation interval. The lower pre-error threshold is set to $n_{limit\downarrow} = 1$, the circuit is operated at a maximum delay of $t_{d,max} = 1.7$ ns and the clock period is $T_{Clk} = 2$ ns

controlled manner by changing the AVS settings. Besides the threshold $n_{limit\downarrow}$, also the number of clock cycles N for the observation interval and/or the length of the pre-error detection window T_{pre} can be changed. Figure 6.5 shows $P_{V_{DD}\downarrow}$ over the parameters N and T_{pre} for a maximum delay of $t_{d,max} = 1.7$ ns. To obtain such a maximum delay, the voltage is reduced to the appropriate value under nominal process and at a temperature of $T = 27$ °C. Note that we chose this maximum delay as at that point, a further reduction of V_{DD} by one step might already lead to setup time violation and consequently the risk of timing errors. Furthermore, the lower pre-error limit is set to $n_{limit\downarrow} = 1$, which is the most conservative setting as the voltage is then only reduced for zero pre-errors during an observation interval.

For the chosen lower pre-error threshold of $n_{limit\downarrow} = 1$, Eq. 6.1 simplifies to

$$P_{V_{DD}\downarrow} = P[n_{pre} = 0] = (1 - P_{pre})^{N} \quad (6.2)$$

The exponential dependence of $P_{V_{DD}\downarrow}$ in Eq. 6.2 can be seen along the N-axis. Along the T_{pre}-axis, the effect of changing pre-error probabilities P_{pre} with the length of the detection window gets visible.[2] Thus, with increasing N and/or T_{pre} the probability $P_{V_{DD}\downarrow}$ decreases. This way, the risk of critical voltage reductions can be minimized.

However, a lower risk of over-aggressive voltage scaling generally reduces the probability for a voltage decrement—also in situations with relaxed timing slack. This in turn reduces the power savings. In other words, the power saving potential can be traded for a lower risk of timing errors. Due to the statistical behavior of the pre-error count, a large N is favorable for the voltage regulation. However, when testing different values for N, our analysis showed that an observation interval longer than $N = 1000$ clock cycles only slightly improves the trade-off between reliability and power saving.

[2]The pre-error probabilities P_{pre} are again extracted from the SPICE simulation results. For a longer pre-error detection window more transitions are detected as pre-error and P_{pre} increases.

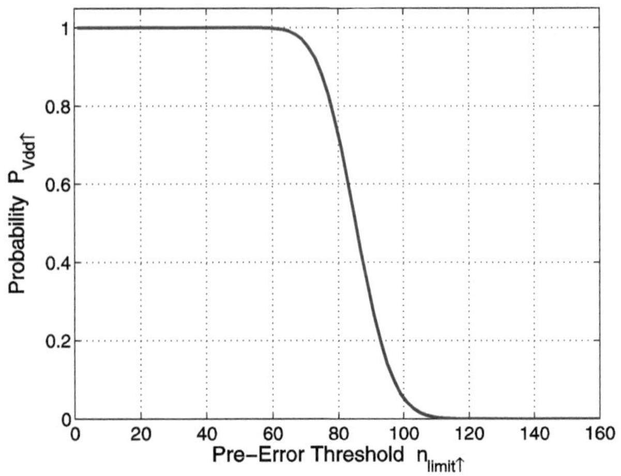

Fig. 6.6 Probability $P_{V_{DD}\uparrow}$ in dependence of the pre-error threshold $n_{limit\uparrow}$ for a timing slack of zero ($P_{pre} = 0.086$). The observation interval has a length of $N = 1000$ clock cycles

We also found out that for the analyzed multiplier circuit, a minimum detection window length of around $T_{pre} = 500$ ps is necessary to inhibit over-aggressive voltage scaling and ensure a low timing failure rate. This corresponds to 25 % of the clock period $T_{Clk} = 2$ ns. Our evaluations of the AVS system, described closely in the next chapter, show that a detection window length of 20 % to 30 % of the clock cycle T_{Clk} finally results in error rates around 1E-6 to 1E-12.

So far, we focused on the reduction of the supply voltage. However, in a situation where the circuit slows down (e.g. due to aging) the voltage has to be raised again. To decide when to increase V_{DD}, we again use the pre-error count as a measure. If the pre-error count during an observation interval is larger than a defined threshold $n_{limit\uparrow}$, the voltage is incremented. The probability $P_{V_{DD}\uparrow}$ for this event reads as

$$\begin{aligned} P[n_{pre} > n_{limit\uparrow}] &= P_{V_{DD}\uparrow} \\ &= 1 - \sum_{n_{pre}=0}^{n_{limit\uparrow}} \binom{N}{n_{pre}} \cdot (P_{pre})^{n_{pre}} \cdot (1 - P_{pre})^{N-n_{pre}} \end{aligned} \tag{6.3}$$

At the latest for a timing slack of zero ($T_{slack} = 0$), the voltage should be increased. Otherwise, a further slowing down of the circuit (e.g. due to further aging) results in possible errors. With the example settings from before ($T_{pre} = 600$ ps and $N = 1000$), the pre-error rate equals $P_{pre} = 0.086$ for $T_{slack} = 0$. This results in a mean pre-error count of $\overline{n_{pre}} = P_{pre} \cdot 1000 = 86$. Since the pre-error count during an observation interval is stochastic, it will vary around this mean. The probability $P[n_{pre} > n_{limit\uparrow}] = P_{V_{DD}\uparrow}$, that the pre-error count is above the upper threshold $n_{limit\uparrow}$, and hence the voltage is increased, is depicted in Fig. 6.6.

By setting the pre-error limit $n_{limit\uparrow}$, we can regulate the likelihood for increasing the supply voltage. If we choose the threshold to $n_{limit\uparrow} = 74$, the voltage will be increased with a probability of approximately 90 %. To get a probability of $P_{V_{DD}\uparrow} =$ 99 % we have to reduce the upper limit to 65.

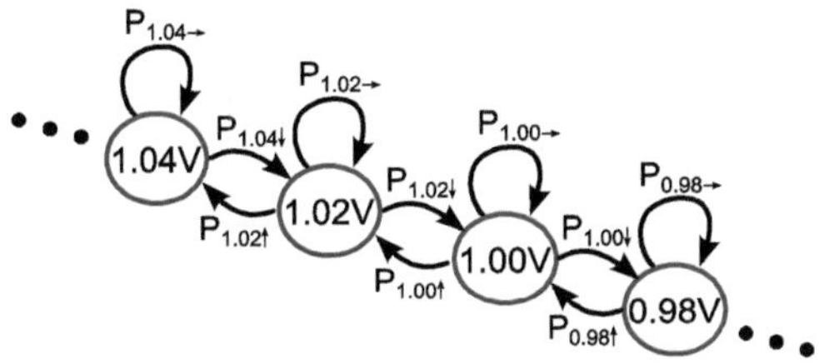

Fig. 6.7 Markov chain with transition probabilities for the individual voltage levels as model for the Pre-Error AVS system

Besides voltage decrements and increments, for a pre-error count between the lower and upper threshold the voltage is maintained. The corresponding probability $P_{V_{DD}\rightarrow}$ for this event reads as

$$P[n_{limit\downarrow} \leq n_{pre} \leq n_{limit\uparrow}]$$
$$= P_{V_{DD}\rightarrow} = \sum_{n_{pre}=n_{limit\downarrow}}^{n_{limit\uparrow}} \binom{N}{n_{pre}} \cdot (P_{pre})^{n_{pre}} \cdot (1 - P_{pre})^{N-n_{pre}} \tag{6.4}$$

For each voltage level, the supply can either be reduced, decreased or maintained. Consequently, the three probabilities sum up to one

$$P_{V_{DD}\downarrow} + P_{V_{DD}\uparrow} + P_{V_{DD}\rightarrow} = 1 \tag{6.5}$$

As the voltage is reduced, increased or maintained only with a specific probability, we propose a Markov chain to model the Pre-Error AVS system [43]. The Markov chain is depicted as directed graph in Fig. 6.7.

The arrows in Fig. 6.7 denote the transition probabilities, i.e. the probabilities to reduce, increase or maintain the voltage. In this example, the used voltage step size is $\Delta V_{DD} = 20$ mV. Such an accuracy is state-of-the-art in typical digital circuits with a low-dropout (LDO) regulator as on-chip power supply unit [47]. The lower the voltage, the smaller the timing slack and consequently the lower the probability $P_{V_{DD}\downarrow}$ to further reduce the voltage and vice versa. The Markov chain, as depicted in Fig. 6.7, is characterized by the corresponding Markov or transition matrix $\boldsymbol{P}$, containing the transition probabilities as elements. With a maximum voltage of $V_{DD} = 1.20$ V the Markov matrix reads as

$$\boldsymbol{P} = \begin{pmatrix} P_{1.20\rightarrow} & P_{1.20\downarrow} & 0 & 0 & 0 & 0 & \cdots \\ P_{1.18\uparrow} & P_{1.18\rightarrow} & P_{1.18\downarrow} & 0 & 0 & 0 & \cdots \\ 0 & P_{1.16\uparrow} & P_{1.16\rightarrow} & P_{1.16\downarrow} & 0 & 0 & \cdots \\ 0 & 0 & P_{1.14\uparrow} & P_{1.14\rightarrow} & P_{1.14\downarrow} & 0 & \cdots \\ \ddots & \ddots & \ddots & \ddots & \ddots & \ddots & \end{pmatrix} \tag{6.6}$$

For a fixed operating condition, the matrix is stationary, i.e. its elements or transition probabilities, respectively, are constant. The corresponding stationary (steady-state) probability vector $\boldsymbol{\pi}$ contains the probabilities $P_{V_{DD}}$ to be at a certain voltage level V_{DD}

$$\boldsymbol{\pi} = \left[\, P_{1.20}\; P_{1.18}\; P_{1.16}\; P_{1.14} \cdots \right] \tag{6.7}$$

The row vector $\boldsymbol{\pi}$ is defined as the left eigenvector of the Markov matrix $\boldsymbol{P}$, associated with eigenvalue one

$$\boldsymbol{\pi} \boldsymbol{P} = \boldsymbol{\pi} \tag{6.8}$$

This means $\boldsymbol{\pi}$ is invariant under application of the transition matrix $\boldsymbol{P}$. When starting the Pre-Error AVS at any voltage level, it converges to the state defined by $\boldsymbol{\pi}$ after a short time. For a fixed operating condition, the probability to be at a certain voltage level remains constant afterward. For changing operating conditions, e.g. for a temperature change, the Markov matrix and thus the probability vector automatically adapt to the new condition. For a fixed operating condition, the AVS system stays well within a narrow voltage range and typically only switches between two voltage levels (see also Sect. 6.3).

6.2.2 Power Saving

The total power consumption of digital circuits consists of the dynamic or active power P_{dyn} and the leakage or static power P_{leak} and can be described by the following formula [65]

$$P_{total} = \underbrace{\alpha \cdot f \cdot C_{switch} \cdot V_{DD}^2}_{P_{dyn}} + \underbrace{V_{DD} \cdot I_{leak}(V_{DD})}_{P_{leak}} \tag{6.9}$$

where α is the switching activity of the circuit, f the clock frequency and C_{switch} the load capacitance of the switching gates. Due to the quadratic voltage dependence of the dissipated dynamic power P_{dyn}, adaptive voltage scaling can achieve considerable savings in dynamic power. Since the leakage current I_{leak} is exponentially dependent on V_{DD}, adapting the supply voltage also significantly reduces the amperage flowing through the circuit in standby mode.[3] Therefore, Pre-Error AVS saves dynamic and leakage power at the same time.

As the voltage switches between different voltage levels in Pre-Error AVS systems, the time spent at the different voltage values has to be considered when evaluating the power consumption. As the time duration at a voltage level is proportional to the probability of that voltage level, the average dynamic and leakage power reads as

$$\overline{P_{dyn}} = \sum_{V_{DD}} p_{V_{DD}} \cdot P_{dyn,V_{DD}} \tag{6.10}$$

and

$$\overline{P_{leak}} = \sum_{V_{DD}} p_{V_{DD}} \cdot P_{leak,V_{DD}} \tag{6.11}$$

[3] The exponential dependence of the sub-threshold leakage current on the drain-source voltage results from drain-induced barrier lowering (DIBL) in short channel devices [66].

where $P_{dyn,V_{DD}}$ and $P_{leak,V_{DD}}$ denote the dynamic and leakage power at a certain voltage level. As an example, $P_{dyn,1.00}$ and $P_{leak,1.00}$ give the power consumptions at $V_{DD} = 1.00$ V. The probabilities $P_{V_{DD}}$ give the likelihood to be at the corresponding voltage levels.

6.2.3 Timing Error Rate

Since the pre-error count is based upon a stochastic process, an overcritical reduction can never be completely ruled out. Besides the power saving potential it is hence essential to quantify the risk of timing errors. Therefore, we have to multiply the error probability for each voltage level $P_{err,V_{DD}}$ with the probability $P_{V_{DD}}$ to be at that voltage. The error rate is given by summing up all products

$$P_{err} = \sum_{V_{DD}} P_{V_{DD}} \cdot P_{err,V_{DD}} \tag{6.12}$$

For the error probabilities $P_{err,V_{DD}}$, timing errors at all outputs have to be considered—no matter if they are equipped with a Pre-Error flip-flop or not. Timing errors occur for delays violating the setup time constraint. The setup time is the minimum amount of time before the triggering clock edge, that the data signal has to be stable in order to be sampled correctly.[4]

For reduced data-to-clock times the clock-to-output (clock-to-Q) delay increases until a point at which the flip-flop fails to latch the input data, see Fig. 6.8. Due to the smooth transition between data-to-clock times with relaxed clock-to-output delay and the point at which the flip-flop fails latching, there exist different definitions for the setup time:

- The data-to-clock time, resulting in a 5 % degradation of the relaxed clock-to-output delay (shown in Fig. 6.8)
- The point at which the clock-to-output versus data-to-clock characteristic has a slope of minus one
- The data-to-clock time with minimum data-to-output delay

We compared the three definitions by applying transistor level simulations and finally chose the 5 %-criterion, as it is the most conservative setup-time definition for the used Transmission-Gate Master-Slave (TGMS) flip-flops.

Regarding the error rate of the AVS system, transitions violating the determined setup time are counted as errors. However, the setup time is not a constant value but changes with the operating condition of the circuit, e.g. it increases for reduced supply voltages, see Fig. 6.8.

---

[4]To ensure proper sampling, the incoming data also has to be kept stable for a minimum amount of time after the triggering edge of the clock (hold time constraint). However, in contrast to the setup-time, the hold-time constraint is relaxed when the supply voltage is scaled. Therefore, the hold time is not considered in the following.

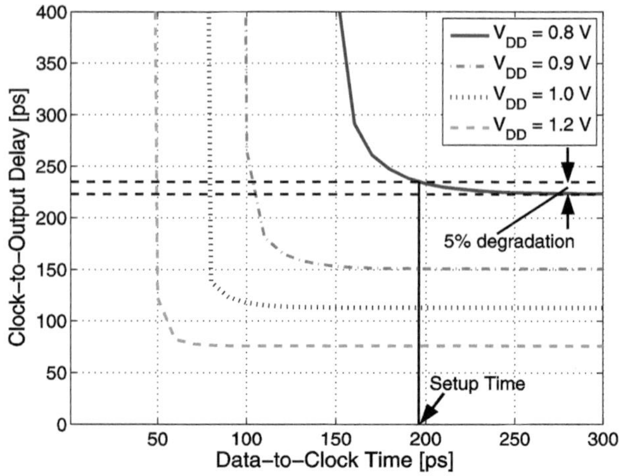

Fig. 6.8 Simulated dependence between data-to-clock time and clock-to-output delay for the used Transmission-Gate Master-Slave (TGMS) flip-flop. The setup time is defined as the data-to-clock time, resulting in a 5 % degradation of the relaxed clock-to-output delay

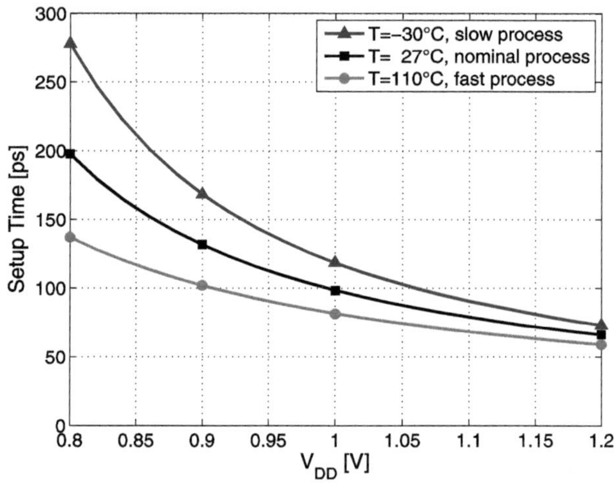

Fig. 6.9 Simulated setup times of the used transmission-gate master-slave flip-flop over the supply voltage V_{DD} for fast, slow and nominal corner

To account for the described setup time fluctuations, we simulated the setup times over the whole voltage range for slow, nominal and fast corner. The resulting setup times, shown in Fig. 6.9, are included into the Markov model to determine the error rate correctly at each voltage level.

Besides global changes, the setup time is also affected by local fluctuations. This is illustrated in Fig. 6.10, where the effect of local process variations on the setup time is illustrated for different voltages. The shown histograms are obtained by Monte-Carlo SPICE simulations. For lower supply voltages the setup time distribution widens, as the flip-flop gets more sensitive to variations.

In order to consider the local setup time variations within the Markov model, we approximate the simulated distributions in Fig. 6.10 by normal distributions. The width of the normal distributions is given by the corresponding standard deviations, which we determined for the different voltage levels. For the used flip-flop,

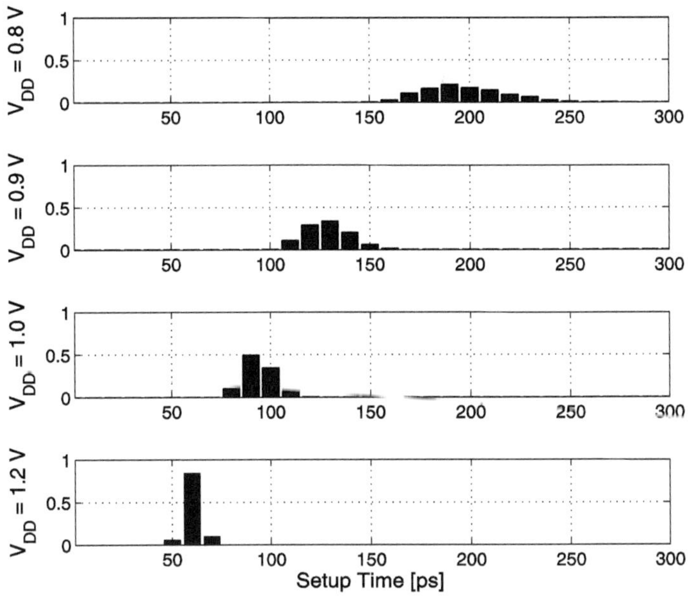

Fig. 6.10 Setup-time distributions of the TGMS flip-flop under local process variations (nominal process, $T = 27\,°C$). The mean value of the setup-time increases for lower supply voltages and the distribution widens

designed in a 65 nm low-power technology, the standard deviation increases from $\sigma_{Setup,1.2\,V} = 3.5$ ps at $V_{DD} = 1.2$ V to $\sigma_{Setup,0.8\,V} = 20$ ps at 0.8 V. By including the obtained distributions into the Markov model, we are able run Monte-Carlo simulations of the modeled AVS system. Therefore, the normal distributions with the corresponding σ-values are used to assign an individual setup time to each flip-flop. For these setup times, the resulting error probabilities $P_{err,V_{DD}}$ are then determined for each voltage level. When including these error probabilities into Eq. 6.12, the overall error rate for each Monte-Carlo run can be calculated.

6.3 Stability of the Control Loop

Based on the pre-error count, the AVS control unit decides after every observation interval—consisting of N active clock cycles—whether the supply voltage V_{DD} is changed. In case of a voltage step (downwards or upwards), the binary control word *Voltage Level* is updated and the voltage regulator (e.g. a LDO) is triggered to adapt its output voltage.

Note that the presented Markov model assumes instantaneous voltage switching, which is an idealization due to the finite response time of the voltage regulator. Figure 6.11 shows a short sequence of voltage adaptation, considering this response time. While the AVS control unit updates the targeted voltage level instantly, the

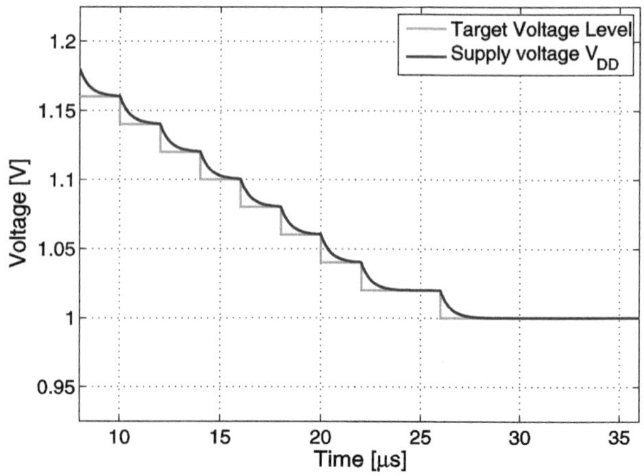

Fig. 6.11 The AVS control unit triggers the voltage adaptation by updating the binary control word *Voltage Level*. The supply voltage V_{DD} follows the targeted voltage level dependent on the time constant of the voltage regulator

voltage regulator follows only with a certain delay dependent on its time constant. In this example the voltage regulator has a time constant of $\tau = 0.5$ µs (which is a typical value for state-of-the-art LDOs [67]) and the observation interval has a length of $N = 1000$ clock cycles. With a clock period of $T_{Clk} = 2$ ns, one observation intervals takes $N \cdot T_{Clk} = 2$ µs, assuming an activity rate of $\alpha = 100$ %. For lower activity rates the time span is accordingly longer and is computed by $N \cdot T_{Clk}/\alpha$.

If the time constant of the voltage regulator is short compared to the time span of the observation interval, the switching behavior resembles instantaneous voltage steps. In this case the Markov model very well reflects the reality. For longer time constants, the system more and more deviates from the ideal case, as the supply voltage follows the target voltage only slowly. This results in a continuous voltage curve and the digital circuit operates at voltages in between the discrete ones (see Fig. 6.11). Consequently, the pre-error rate also changes continuously and is different in every single clock cycle.

To account for the continuous voltage changes, we built a Simulink model which acts like the Markov model, but interpolates the pre-error rate in between the discrete voltage levels. Therefore, the Simulink model observes the present supply voltage and considers the present pre-error rate individually for every clock cycle. Note that this time-based Simulink model has a much longer computation time compared to the state-based Markov model.

With the Simulink model we analyzed the voltage adaptation for different time constants τ of the voltage regulator ($\tau = 0$ µs, $\tau = 1$ µs, $\tau = 4$ µs and $\tau = 12$ µs). The length of the observation interval was kept constant at 2 µs. The results are illustrated in Fig. 6.12. For the ideal case ($\tau = 0$), the voltage regulator directly follows the target voltage. In contrast, if the time constant is large compared to the length of the observation interval, the target voltage begins to oscillate. This effect can be explained as follows: When the target voltage is reduced (due to a pre-error count below the lower threshold $n_{limit\downarrow}$) the supply voltage only follows slowly. Thus, the voltage during the next observation interval is almost as high as in the

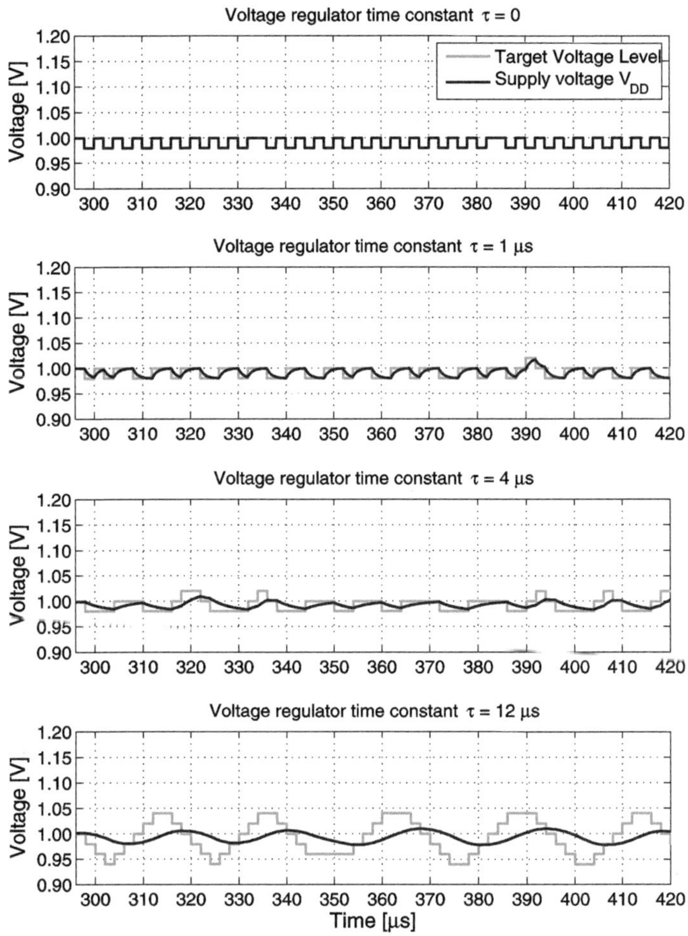

Fig. 6.12 Voltage adaptation for various time constants τ of the voltage regulator. For time constants large compared to the length of the observation interval (2 μs), the target voltage level begins to oscillate. The supply voltage (output of the voltage regulator) stays however still within the same narrow voltage range

previous one. Accordingly, the probability for a low pre-error count is almost the same than before. This in turn makes a further reduction of the target voltage level very likely. The same effect arises when the voltage is increased. As a consequence, the target voltage oscillates. This oscillation, on the other hand, is compensated again by the time constant of the voltage regulator and the supply voltage stays still within the same narrow voltage range as in the ideal case.

In other words, the slew rate of the voltage regulator causes target voltage oscillations and at the same time damps them. Even for a time constant of $\tau = 12\ \mu s$, the voltage stays within the same voltage range as in the ideal case. Detailed analysis also showed, that the error rate as well as the power saving remain unchanged for the different τ-values. Accordingly, the Markov model gives the same results as the Simulink model, while being much less compute intensive.

Note that in Fig. 6.12, the supply voltage switches almost equally between two voltage levels ($V_{DD} = 1.00$ V and $V_{DD} = 0.98$ V). Such a constellation is very seldom, but was chosen to illustrate the oscillation phenom. For a fixed operating condition, the voltage typically remains mainly at one level and changes rather infrequently. Also note that the mentioned oscillations—though they do not harm the voltage adaptation—can be reduced by enlarging the observation interval.

Chapter 7
Evaluation of the Pre-Error AVS Approach

To evaluate the power saving potential of the Pre-Error AVS technique we use the Markov chain model, described in the previous chapter. The Markov chain describes the switching between the discrete output levels of the voltage regulator and thus represents the whole control system. With this model, the resulting probabilities of being at the individual voltage levels can be determined very fast. Moreover, the effect of global as well as local variations on the voltage control can be analyzed very accurately and efficiently. In the following we will also cover the power overhead introduced by the necessary AVS circuitry consisting of the Pre-Error flip-flops and the AVS control unit. As a proof of concept, the application of Pre-Error AVS to an image processing circuit is finally shown.

7.1 Power Saving Potential and Reliability

Due to the Law of Large Numbers (LLN) [68], statistical outliers of the pre-error count become very unlikely for long observation intervals. Therefore, a long observation interval is advantageous to the reliability of the Pre-Error AVS system. Observation intervals longer than $N = 1000$ cycles, however, only slightly improve the voltage adaptation to fixed operating conditions, while increasing area and power overhead of the AVS control. Longer observation intervals also slow down the reaction time when the operating condition of the circuit changes. Consequently, for the following results, N is fixed to 1000. By tuning the other AVS settings (T_{pre}, $n_{limit\downarrow}$ and $n_{limit\uparrow}$) a desired error probability can be achieved. For a fixed setting, the voltage is then adapted automatically dependent on the actual circuit speed. This way the error rate is kept constant under PVT variations and aging.

7.1.1 Voltage Adaptation Under Global Variations

Process variations due to imperfections during the manufacturing process remain unchanged afterward and are automatically covered by the Pre-Error AVS system.

M. Wirnshofer, *Variation-Aware Adaptive Voltage Scaling for Digital CMOS Circuits*, Springer Series in Advanced Microelectronics 41, DOI 10.1007/978-94-007-6196-4_7,

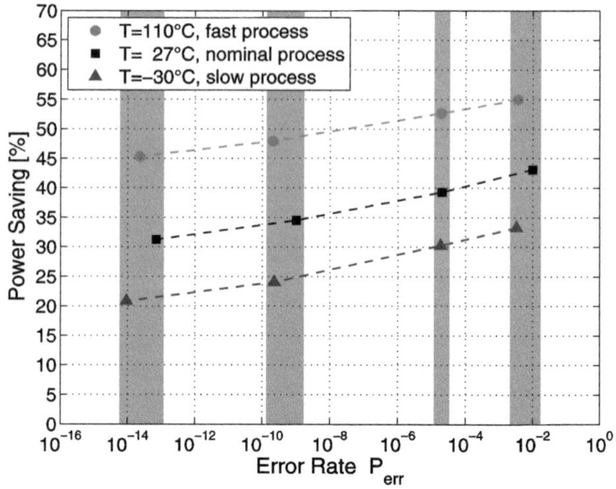

Fig. 7.1 Correlation between error rate P_{err} and dynamic power saving for a 65 nm 16-bit multiplier when applying AVS with Static Pre-Error flip-flops. The power savings are shown for fast, nominal and slow corner. The user can adjust the different error rates in a controlled manner by changing the AVS settings

The on-chip temperature typically changes with a time constant of milliseconds to seconds [22]. Thus, temperature variations are slow compared to the response time of the control loop and are also easily covered. The same is true for long-term aging effects in digital circuits. The AVS also compensates for a too conservative timing estimate of the synthesis tools (exemplified by Fig. 6.1 on page 46) as well as for inaccuracies of the voltage regulator, e.g. due to an offset in the bandgap reference.

For fast changing effects, such as voltage drops due to activation of neighboring logic blocks, we considered the same amount of safety margin in the worst-case design and the AVS system. In practice, the synthesis is performed at nominal voltage ($V_{DD} = 1.20$ V) minus a safety margin of 20 mV and the error estimation for the AVS scheme is also done 20 mV below the actually applied voltage level. Choosing a safety margin of 20 mV for power noise effects is only an example. We can arbitrarily change this margin in our Markov model and this way make a fair comparisons between a worst-case design with a specified safety margin and the corresponding AVS design.

To analyze the power saving potential of the Pre-Error AVS concept, we use a synthesized 65 nm 16-bit multiplier circuit. The clock frequency of the multiplier is $f = 500$ MHz ($T_{Clk} = 2$ ns) and we equipped the three most critical outputs with Static Pre-Error flip-flops, i.e. 9.4 % of all 32 outputs. We evaluated different AVS settings and Fig. 7.1 illustrates the relation between the achieved dynamic power savings and the corresponding timing error rates. The power savings always refer to the same worst-case design, which ensures proper working for slow process, a voltage of $V_{DD} = 1.18$ V and the whole temperature range from -30 °C to 110 °C. The power consumption for the worst-case design and the dynamic power consumption at each voltage level $P_{dyn,V_{DD}}$ are determined by SPICE simulations. The $P_{dyn,V_{DD}}$-values are fed into Eq. 6.10 to obtain the resulting power savings for the Pre-Error AVS approach.

The colored regions in Fig. 7.1 mark fixed AVS settings. For the different regions, the pre-error detection window length T_{pre} is gradually decreased from the leftmost ($T_{pre} = 650$ ps) to the rightmost region ($T_{pre} = 275$ ps). The values in between are $T_{pre} = 550$ ps and $T_{pre} = 400$ ps. The pre-error thresholds are the same for all regions and set to $n_{limit\downarrow} = 1$ and $n_{limit\uparrow} = 20$, respectively. For the leftmost region, the settings are chosen to obtain a timing error rate of $P_{err,nom} = 1\text{E-}13$ in the nominal case ($T = 27$ °C, nominal process). This results in dynamic power savings of 31 % compared to the worst-case guard-banding design. For a shorter detection window length, the error rate as well as the power saving increases. For an error rate of 1E-2 (rightmost region), the dynamic power is reduced by 43 %.

Thus, by changing the AVS settings (in this example T_{pre} is changed), the error rate can be regulated in a wide range and the circuit can operate at peak efficiency for the accepted error rate of the application. Note that the same error rate can be achieved with different combinations of T_{pre}, $n_{limit\downarrow}$ and $n_{limit\uparrow}$. However, for the power saving it is advantageous to target the tail of the delay distributions (see Fig. 6.3 on page 49) by applying short detection windows in combination with low pre-error thresholds, rather than long detection windows in combination with higher thresholds. The reason therefore can be explained as follows: When only the tail of the distribution lies inside the pre-error detection window, a change in operating condition strongly affects the pre-error rate. This change of the pre-error rate gets less steep if a larger portion of the distribution is inside the detection window. For the Pre-Error AVS system, a higher slope of the pre-error rate leads to a more reliable distinction between relaxed and critical operating conditions. Thus, lower error rates are achieved for the same power savings.

For the fast corner ($T = 110$ °C, fast process),[1] the switching speed of the circuit is higher and the AVS system adapts the voltage therefore to lower V_{DD}-levels, which leads to higher power savings. But even for the slow corner, there is still considerable power saving. This is due to the fact, that the AVS system exploits the unused timing margin produced by the conservative timing estimation of the synthesis tool.

Under global variations, the Pre-Error AVS scheme keeps the error rate almost constant by tuning the supply voltage accordingly. This can be seen, when comparing slow and fast corner in Fig. 7.1. The reason is that global fluctuations hardly shift the detection window length of the Static Pre-Error flip-flop and affect all multiplier outputs similarly. Note that also for the Dynamic Pre-Error flip-flop and the Crystal-ball flip-flop the error rate is almost unchanged under global variations. Especially for the Crystal-ball flip-flop that seems surprising, as the delay element, defining the detection window, is heavily affected by global variations. However, the AVS system balances this effect by adapting the voltage accordingly. This way the actual propagation time through the delay element varies only slightly. For example, if the digital logic and consequently also the delay element is slowed down due to changing operating conditions, the voltage is increased and the detection window length (propagation time through the delay element) stays almost constant.

[1] Note that typically the fast corner is at fast process and *low* temperature. In the presented AVS system, however, the voltage is scaled to a point where the circuit is operated at temperature inversion. This means that the effect of decreasing threshold voltage with temperature exceeds the mobility degradation and the circuit speeds up with increased temperature and vice versa.

Table 7.1 Overhead (in %) of active power for the necessary AVS circuitry, when applying the Pre-Error AVS system to the multiplier test circuit. Remember that three of the 32 output flip-flop are replaced by Pre-Error flip-flops

	Crystal-ball	Dynamic	Static
Overhead of the three Pre-Error flip-flops	9.6[a]	9.4	3.9
Overhead due to the AVS control unit	3.6	3.6	3.6
Total power overhead	13.2	13.0	7.5

[a]The values for the Crystal-ball flip-flop are obtained for a nominal detection window length of $T_{pre} = 600$ ps. The power consumption increases for longer detection windows and vice versa

7.1.2 Total Power Savings with Pre-Error AVS

Getting back to the power analysis, Fig. 7.1 shows the gross savings in dynamic power for the digital circuit. This means, the power overhead of the AVS circuitry (Pre-Error flip-flops and AVS control unit) is not included yet in these numbers. To determine the mentioned overhead, we simulated the power consumption of the three different kinds of Pre-Error flip-flops (Static, Dynamic and Crystal-ball) and the AVS control unit by transistor level simulations. The power overheads in typical operation mode are listed in Table 7.1.

The values are given in percent and refer to the power consumption of the worst-case design, which is 1.2 mW. The relative overhead of the AVS control unit is 3.6 %. For digital circuits, that are larger than the rather small multiplier circuit, the relative overhead will even be smaller. The power overhead of the Dynamic Pre-Error flip-flop and the Crystal-ball flip-lop is more than twice the overhead of the Static Pre-Error flip-flop. This is due to the high transistor count and the delay element, respectively (see also Chap. 5). To get the net power savings, the total power overhead has to be subtracted from the power savings shown in Fig. 7.1. If the Static Pre-Error flip-flop is chosen, the net savings result in 35.6 % to 23.7 % (gross savings: 43.1 % to 31.2 %).

So far we only discussed the active power consumption, but when scaling the voltage, also the leakage power is reduced significantly. Dependent on the AVS setting (error rate) we get a gross reduction of leakage power by 44 % to 57 % for the multiplier circuit in nominal case. With a simulated leakage power overhead of approximately 10 % when using Static Pre-Error flip-flops, the net savings in leakage power are hence 34 % to 47 %. For the Crystal-ball and the Dynamic Pre-Error flip-flop, the leakage currents are larger due to the higher transistor count, which results in around five percent less savings (29 % to 42 %).

For the Static and Dynamic Pre-Error flip-flop, it is also necessary to define the detection window by adjusting the duty-cycle. Generating this clock signal adds no extra power if the system already comprises a multiphase clock generator, which is the case for many microprocessors and other sophisticated circuits [49]. If the pulse-width control has to be implemented additionally, this has to be included in the power analysis of the AVS system.

Moreover, making an existing LDO capable of changing its output voltage with a specified step size ΔV_{DD} can be achieved without power overhead. Instead of the 1:1 voltage divider in the feedback loop [46], which halves the voltage V_{DD}, we propose to implement a resistor ladder with switches (see Fig. 4.8 on page 30). These switches are driven by the binary coded Voltage Level of the AVS control unit. The necessary switching power is already included in the power overhead of the AVS control unit. Thus, regarding the Pre-Error AVS system, no extra power for the voltage regulator has to be considered.

7.1.3 Voltage Adaptation Under Local Variations

Besides the effect of global variations, we also analyzed the impact of local variations on the voltage adaptation. Local variations result in an error uncertainty ΔP_{err}, which is illustrated by the cloud diagram in Fig. 7.2. For the shown results, local process variations of the Static Pre-Error flip-flops (including the clock-tree) and the multiplier circuit itself are considered.

Therefore, we first determine the effect of local variations on the detection window of a single Pre-Error flip-flop. This is done by Monte-Carlo SPICE simulations (see also Fig. 5.12). Monte-Carlo SPICE simulations are also applied to determine the effect of local variations on the output delays of the multiplier circuit. The simulated statistics (standard deviations) were included in the Markov model to obtain the influence of local variations on the overall system.

The Markov model uses these statistics to generate an independent detection window deviation for every Pre-Error flip-flop. This is done by a random generator, considering the distribution of the detection window length. Also for every output, a random generator determines a delay shift dependent on the standard deviation of the corresponding output. Taking the individual detection window length in combination with the shifted delay histogram at each output, the Markov model determines the new pre-error probability P_{pre} for every voltage level. From the shifted delay histograms also the resulting timing error rates $P_{err,V_{DD}}$ are obtained.

By our two step approach (first simulating on SPICE level, then including the resulting statistics into the Markov model), the effect of local variations can be analyzed very efficiently. Running Monte-Carlo SPICE simulations for the whole system would be computationally infeasible and hardly increases the accuracy of the results.

7.1.4 Role of In-situ Delay Monitors

In the last subsection, we analyzed the effect of local detection window variations in combination with local fluctuations of the multiplier circuit. In the following, we focus on the local variations of the Pre-Error flip-flops. Therefore, we reuse our

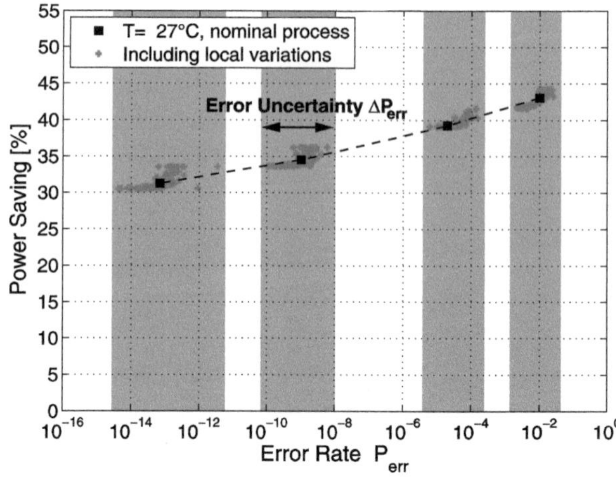

Fig. 7.2 Impact of local process variations of the multiplier circuit and the Static Pre-Error flip-flops (plus clock tree) on the voltage adaptation. The cloud diagram shows the outcome of 100 MC runs when including the variation effects into the Markov model

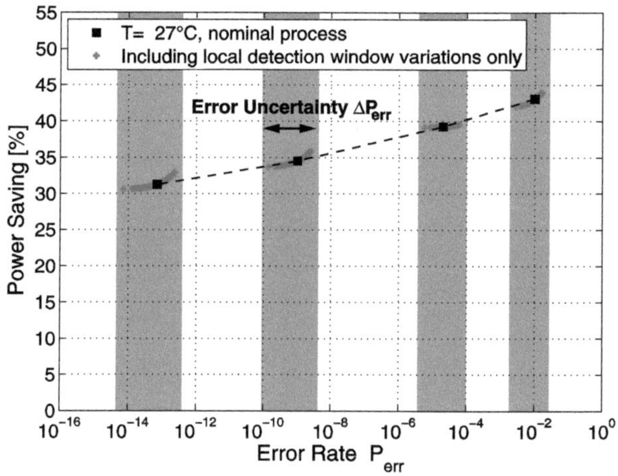

Fig. 7.3 Impact of local detection window variations on the voltage adaptation. The local detection window variations include variations of the Static Pre-Error flip-flops and the clock tree (variations of the multiplier circuit are excluded)

two step simulation approach, which can also analyze the effect of multiplier and detection window variations apart from each other. Figure 7.3 shows the resulting error uncertainty for 100 Monte-Carlo runs when using Static Pre-Error flip-flops and excluding variations of the multiplier circuit.

Compared to Fig. 7.2 (showing the combined variation effect), the obtained error uncertainties are significantly lower. Thus, the combined variation effect (Fig. 7.2) is rather caused by the local variations of the multiplier circuit than the uncertainty of the detection window. In other words, the multiplier variations dominate the fluctuations of the Pre-Error flip-flops and the combined variation effect (Fig. 7.2) looks similar for all three Pre-Error flip-flops.

To compare the different Pre-Error flip-flops, we did the same Monte-Carlo analysis (shown in Fig. 7.3) also for the Crystal-ball and Dynamic Pre-Error flip-flop. As metric for the robustness of the Pre-Error flip-flops we use the normalized error

Table 7.2 Relative error uncertainty for the different types of Pre-Error flip-flops. The uncertainty is obtained for a nominal error rate of 1E-9 when considering local detection window variations

	Crystal-ball	Dynamic	Static
Normalized error uncertainty for a nominal error rate of 1E-9	1.6	4.8	2.3
3σ deviation of T_{pre} under local variations at 0.8 V (see Table 5.1)	69 ps	91 ps	49 ps

uncertainty, which is obtained by dividing the error uncertainty ΔP_{err} by the nominal error rate $P_{err,nom}$. Table 7.2 summarizes the relative error uncertainties for a nominal error rate of $P_{err,nom} = 1\text{E-}9$.

With a value of 2.3, the normalized error uncertainty for the Static Pre-Error flip-flop is considerably lower compared to the Dynamic Pre-Error flip-flop (4.8). This is reasonable as the detection window of the Static Pre-Error flip-flop is more robust against local variations than the detection window of the dynamic one. (The three times standard deviation for the Static Pre-Error flip-flop reads as $3\sigma_{0.8\text{ V}} = 49$ ps compared to $3\sigma_{0.8\text{ V}} = 91$ ps for the Dynamic Pre-Error flip-flop, see also Table 5.1.)

The Static Pre-Error flip-flop ($3\sigma_{0.8\text{ V}} = 49$ ps) is also more robust against local variations compared to the Crystal-ball flip-flop ($3\sigma_{0.8\text{ V}} = 69$ ps). Nevertheless, the relative error uncertainty of 1.6 for the Crystal-ball flip-flop is the lowest of three values.

The reason for this surprising result is the strong voltage dependence of the delay element in the Crystal-ball flip-flop. When the voltage is decreased, not only the logic slows down, but also the delay element. This enlarges the detection window and additionally increases the pre-error rate. This makes a subsequent voltage increment very likely. The same is true vice versa. Thus, the strong voltage dependence of the Crystal-ball flip-flop stabilizes the adapted voltage around one level and conceals the effect of local detection window variations.

7.2 Application of Pre-Error AVS for an Image Processing Circuit

Digital image and video processing is a key task in many areas, like multimedia, medicine or robotics. Due to its high relevance, we chose this field of application to analyze the suitability of the Pre-Error AVS concept. As test vehicle we designed a Discrete Cosine Transform (DCT) circuit, which is used in current image and video coding standards like JPEG, MPEG, H.261 or H.263 [69, 70]. Therefore, the DCT was first described with VHDL and then synthesized as semi-custom design. The resulting worst-case design serves as reference to analyze the power saving potential of our Pre-Error AVS concept.

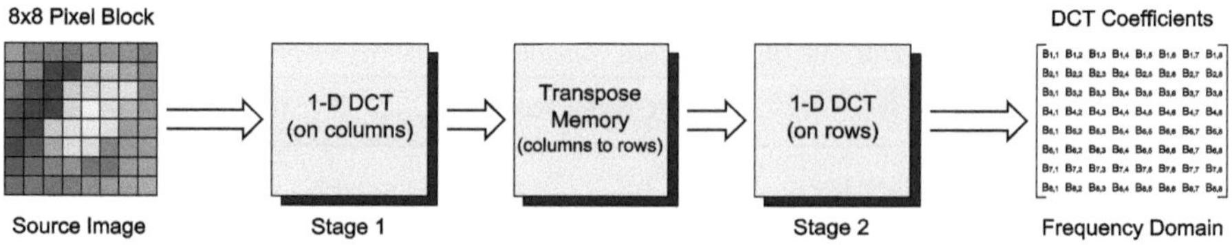

Fig. 7.4 Block diagram of the 2-D DCT, implemented by two 1-D DCT operations along the two dimensions of the image

7.2.1 Discrete Cosine Transform (DCT) Circuit

The discrete cosine transform (DCT) expresses a discrete time domain signal by a weighted sum of cosine functions and this way transforms it into the frequency domain. The DCT is similar to the discrete Fourier transform (DFT), but uses only cosine functions instead of sine *and* cosine functions to represent the signal. As the DCT can concentrate the signal information on less coefficients than the DFT, it is the method of choice in many data compression algorithms such as MP3, JPEG or MPEG.

For MP3, the one dimensional (1-D) DCT is used, whereas image and video compression requires the 2-D DCT. As the 2-D DCT is a separable linear transformation, it can be computed by applying a 1-D DCT along the one dimension, followed by a second 1-D DCT along the other dimension of an image. This property is often also exploited when implementing the 2-D DCT on hardware. Figure 7.4 visualizes the working principle of such an implementation. The pixel information of an image block is transformed along the columns by the first 1-D DCT (stage 1). The transposed information is subsequently processed along the rows by to the second 1-D DCT (stage 2), which outputs the resulting DCT coefficients.[2]

Our implementation of the DCT test vehicle is also based on the described two stage concept. For more information on the details of the implementation we refer to [71]. When designing the DCT, we used the classical semi-custom approach and described the DCT first on VHDL level. Afterward, it was synthesized in a 65 nm low-power technology with industrial EDA tools. The clock frequency is set to $T_{Clk} = 2$ ns ($f = 500$ MHz) and the synthesis is performed using traditional worst-case guard-banding. The worst-case is defined at slow corner and a temperature range from $T = -30$ °C to $T = 110$ °C. The nominal operating voltage is set to $V_{DD,nom} = 1.2$ V with a voltage safety margin of 20 mV, i.e. proper working is ensured down to 1.18 V. As for the multiplier circuit, the power consumption of the synthesized worst-case design serves as reference when evaluating the power saving potential of our Pre-Error AVS approach.

[2]Further information on the working principle of the 2-D DCT can be found in the Appendix.

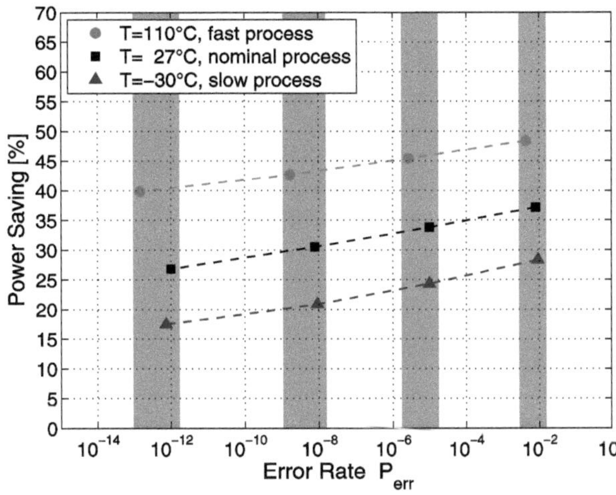

Fig. 7.5 Correlation between timing error rate P_{err} and dynamic power saving of the AVS approach for different corners. The error rate can be controlled by tuning the AVS settings. In this case by changing the pre-error detection window length from $T_{pre} = 550$ ps to $T_{pre} = 250$ ps in steps of 100 ps for the four colored regions. The observation interval is fixed to $N = 1000$, the pre-error limits are $n_{limit\downarrow} = 1$ and $n_{limit\uparrow} = 20$

7.2.2 Power and Performance Analysis

To evaluate the application of the Pre-Error AVS concept to the DCT circuit [45], we equipped the 52 most critical outputs (found by SPICE simulations with random input pattern) with Static Pre-Error flip-flops. This corresponds to 5 % of all 1045 flip-flops, which is a typical coverage when implementing in-situ delay monitors [10]. The length of the observation interval is set to $N = 1000$, i.e. after every 1000 active clock cycles the AVS control unit decides whether the voltage is increased, decreased or held constant. The pre-error limits are set to $n_{limit\downarrow} = 1$ and $n_{limit\uparrow} = 20$. The length T_{pre} of the pre-error detection window is varied from 550 ps to 250 ps to adjust the timing error rate. The relation between error rate and dynamic power saving is shown in Fig. 7.5. The four colored regions denote the different settings for the pre-error detection window length. For the leftmost region it is set to $T_{pre} = 550$ ps and is reduced for the other regions to 450 ps, 350 ps and 250 ps. The results for slow and fast corner show that the error rate is kept almost constant under global variations.

In the nominal case, the dynamic power is reduced by 27 % to 37 % for error rates between 1E-12 and 1E-2. Thus, for similar error rates the power savings are slightly lower than for the multiplier circuit. The reason therefore lies in synthesis of the worst-case design. When analyzing the timing of the worst-case design by SPICE simulations, we discovered that the DCT circuit is slightly less conservative than the multiplier circuit. In case of the DCT, the timing estimation of the STA is somewhat closer to the real circuit behavior and hence less timing margin is produced by the synthesis tools. However, the results of the DCT circuit are very similar to the ones of the multiplier and thus confirm the power saving potential of our AVS scheme.

Table 7.3 Key characteristics of the Pre-Error AVS test circuit

Key metrics of the DCT circuit	
Technology node	65 nm (low power)
Voltage range	1.2 V to 0.8 V
Clock frequency	500 MHz
Power consumption of worst-case design	19.9 mW
Total number of logic gates	11444
Total number of flip-flops	1045
Number of Pre-Error flip-flops	52 (≘5 %)
AVS settings	
Observation interval	$N = 1000$
Lower threshold limit	$n_{limit\downarrow} = 1$
Upper threshold limit	$n_{limit\uparrow} = 20$
Pre-Error detection window length	$T_{pre} = 550$ ps to 250 ps
Power and area analysis	
Power consumption of worst-case design	19.9 mW
Timing error rate	1E-12 to 1E-2
Dynamic power saving in nominal case[a]	23 % to 33 %
Leakage power saving in nominal case[a]	34 % to 45 %
Power overhead AVS control unit	43 μW (≘0.2 %)
Power overhead Pre-Error flip-flops (plus OR-trees)	820 μW (≘4.1 %)
Area overhead AVS control unit	0.9 %
Area overhead Pre-Error flip-flops (plus OR-trees)	3.8 %

[a]These numbers already consider the power overhead produced by the AVS circuitry and thus represent the net savings

When considering the power overhead[3] of the AVS control unit (0.2 %) and the in-situ delay monitors plus OR-trees (4.1 %), the net power savings are reduced to 23 %–33 %. Due to the larger system, the relative power overhead of the AVS control unit is considerably lower than for the multiplier circuit (3.6 %) and can almost be neglected. Regarding the area overhead, the in-situ delay monitors (plus OR-trees) add 3.8 % and the AVS control unit adds another 0.9 % in chip-area.

By scaling the supply voltage, also the leakage power is reduced. For the DCT circuit, a net leakage power reduction of 34 % to 45 %, dependent on the error rate (1E-12 to 1E-2) is achieved. This large reduction is due to the exponential dependence of the leakage current on V_{DD}. All key metrics of our AVS test circuit are summarized in Table 7.3.

[3]Note that the power consumption of all parts of the system is simulated by transistor level simulations (SPICE) to obtain accurate results.

(a) Original Image

(b) Error rate: $P_{err} = 10^{-2}$
Power Saving: 33%

(c) Error rate: $P_{err} = 10^{-5}$
Power Saving: 29%

(d) Error rate: $P_{err} = 10^{-8}$
Power saving: 26%

Fig. 7.6 Image quality for different error rates of the DCT circuit. The error rate and thus the image quality is user-defined by means of the AVS settings

So far, we only considered timing error rates in our analysis. For the DCT, however, the resulting image quality is the key metric. Hence, we evaluated the effect of a certain error rate on the computation of the DCT coefficients and finally on the image quality. Therefore, we operated the DCT at different error rates and stored the resulting DCT coefficients. Then we transformed the image back to the spatial domain by the inverse DCT (IDCT), which is performed ideally, i.e. without any error.

The resulting images for different error rates are shown in Fig. 7.6. From the figure it can be seen that for AVS settings with an error rate of 1E-2 the DCT circuit yields only poor image quality. However, for error rates of 1E-5 only few 8×8 pixel blocks per image get corrupted. By further reducing the error rate to 1E-8, errors occur only in one out of 250 pictures, while still saving 26 % of active power and 34 % of leakage power.

Chapter 8
Conclusion

To boost the power efficiency of digital CMOS circuits, a novel Adaptive Voltage Scaling (AVS) approach is proposed, which is capable of handling the increasing PVT variations as well as aging effects in integrated circuits. This Pre-Error AVS approach reduces unnecessary voltage safety margin by adapting the supply voltage to the actual operating condition of a chip and thereby optimizes power consumption.

In the Pre-Error AVS scheme the timing information (operating condition) of the chip is taken inside the real circuit (in-situ) and measured during normal circuit operation (on-line). Therefore, Pre-Error flip-flops are used, which detect late but still non-erroneous transitions (pre-errors) in critical paths. Due to the on-line monitoring, no interrupts for test-cycles are required, which makes the voltage scheme viable for real-time applications. The gathered timing information is processed by a newly developed AVS control unit, which triggers the voltage regulator to tune the supply voltage. This AVS control unit is completely digital and can thus be synthesized easily in every new technology. Moreover, it is compact and introduces little power and area overhead.

For the AVS control unit, observation intervals with a specified number of clock cycles N are introduced. Furthermore, a lower ($n_{limit\downarrow}$) and upper threshold ($n_{limit\uparrow}$) for the pre-error count (n_{pre}) of each observation interval is defined. With these settings, the control unit decides after every observation interval whether the voltage is decreased (for $n_{pre} < n_{limit\downarrow}$), increased (for $n_{pre} > n_{limit\uparrow}$) or maintained (for $n_{limit\downarrow} \leq n_{pre} \leq n_{limit\uparrow}$).

As mentioned before, the AVS control unit relies on the pre-error rate obtained by the elaborated Pre-Error flip-flops. These enhanced flip-flops have to distinguish precisely between pre-errors (transitions during the pre-error detection window) and regular transitions. Additionally, the defined detection window length T_{pre} must be robust against all kinds of variations. To fulfill these requirements, a new type of Pre-Error flip-flop is introduced. Instead of using a delay-element to define the detection window, this flip-flop exploits the duty-cycle of the clock. In a detailed analysis it is shown that this new Pre-Error flip-flop outperforms state-of-the-art solutions with respect to power and area consumption as well as robustness.

M. Wirnshofer, *Variation-Aware Adaptive Voltage Scaling for Digital CMOS Circuits*, Springer Series in Advanced Microelectronics 41, DOI 10.1007/978-94-007-6196-4_8,

For reliable voltage adaptation, the Pre-Error flip-flop is further improved by adding a transition monitor, which observes the activity of the circuit. This way, only active clock cycles are counted for the observation interval and over-aggressive voltage scaling is avoided.

According to systems theory, the presented AVS scheme forms a closed-loop control with the Pre-Error flip-flops acting as sensors and the AVS control unit together with the voltage regulator acting as controller. As the voltage adaptation is done on-line, the randomness of the input pattern influences the voltage control. To model this stochastic control system, a sophisticated Markov chain was established. This model was the key for analyzing the power saving potential and reliability of the Pre-Error AVS system. The necessary parameters for the Markov chain are determined by extensive transistor level simulations in order to obtain an accurate model. Regarding the stability of the control loop, it is shown that the system's inherent self-regulation avoids unstable behavior.

The power saving potential was investigated thoroughly for a 16-bit multiplier circuit and an image processing circuit. Both designs were synthesized in a 65 nm low-power technology. For the widely used multiplier circuit, the dynamic power is reduced by 36 % to 24 % for timing error rates of 1E-2 to 1E-12, the additional components (Pre-Error flip-flops and the AVS control unit) already included. The implied supply voltage reduction automatically decreases the leakage power by more than one third. It is illustrated how the system can be tuned to obtain different error rates by changing the AVS settings. These are the duration of the observation interval, the pre-error thresholds and the detection window length. This way, reliability (error rate) can be traded for power savings and the circuit operates at peak efficiency for a desired error rate. When evaluating different AVS settings, it was found that the detection window length has to be around 20 % to 30 % of the clock cycle T_{Clk} to achieve error rates of 1E-6 to 1E-12.

With the aforementioned Markov model, also the effect of global and local variations on the voltage adaptation can be analyzed. This analysis shows that the adjusted error rate is hardly affected by global variations as the AVS system accurately compensates them. Under local fluctuations the error rate varies somewhat stronger, but still within a narrow range.

The application of the Pre-Error AVS approach to an image processing circuit demonstrates the suitability of the concept also for larger circuits. It is also shown how the adjusted timing error rate affects the quality of the processed images. The example therefore provides a better understanding of the relevance of timing errors on gate level for the resulting performance on system level. Savings of almost 30 % in active power and about 40 % in leakage power are demonstrated, while barely affecting the image quality (timing error rate: 1E-8 to 1E-5). At the same time, the necessary AVS circuitry (Pre-Error flip-flops and control unit) introduces an area overhead of only 5 %.

In conclusion, it can be stated that the elaborated Pre-Error AVS concept achieves considerable power savings and is applicable for every general purpose digital circuit. This makes it a highly considerable solution to the increasing concerns about power consumption and aging in integrated circuits. Also in the emerging fields of low-voltage and sub or near threshold circuits [72–74], Pre-Error AVS promises great advantages over the worst-case design. At these voltage levels we see increased sensitivity to PVT variations, which demands for extreme guard-bands to ensure reliable operation. A much more favorable solution is intelligent voltage adjustment with the help of Pre-Error AVS.

Appendix

A.1 Mathematical Derivation: Path Delay Under Local Variations

The variance $\sigma_{t,gate}^2$ for the switching time of one gate under local variations is given by

$$\sigma_{t,gate}^2 = \mathrm{E}\left[(t_{gate} - \mu_{t_{gate}})^2\right]$$

where E denotes the expected value or first moment of a random variable. Accordingly, the variance $\sigma_{t,d}^2$ for the path delay t_d reads as

$$\sigma_{t,d}^2 = \mathrm{E}\left[\left(\sum_{i=1}^{n} t_{gate,i} - \mu_{t_{gate,i}}\right)^2\right]$$

For statistically independent variations of the single gate delays we obtain

$$\begin{aligned}\sigma_{t,d}^2 &= \mathrm{E}\left[\sum_{i=1}^{n} (t_{gate,i} - \mu_{t_{gate,i}})^2\right] \\ &= \sum_{i=1}^{n} \mathrm{E}\left[(t_{gate,i} - \mu_{t_{gate,i}})^2\right]\end{aligned}$$

With the simplification of n equal gates, we get

$$\sigma_{t,d}^2 = n \cdot \mathrm{E}\left[(t_{gate} - \mu_{t_{gate}})^2\right] = n \cdot \sigma_{t,gate}^2$$

A.2 Two-Dimensional DCT Transform

The 2-D DCT transforms the spatial information of an image into the frequency domain. The principle of operation is to express the image information by cosine

M. Wirnshofer, *Variation-Aware Adaptive Voltage Scaling for Digital CMOS Circuits*, Springer Series in Advanced Microelectronics 41, DOI 10.1007/978-94-007-6196-4,

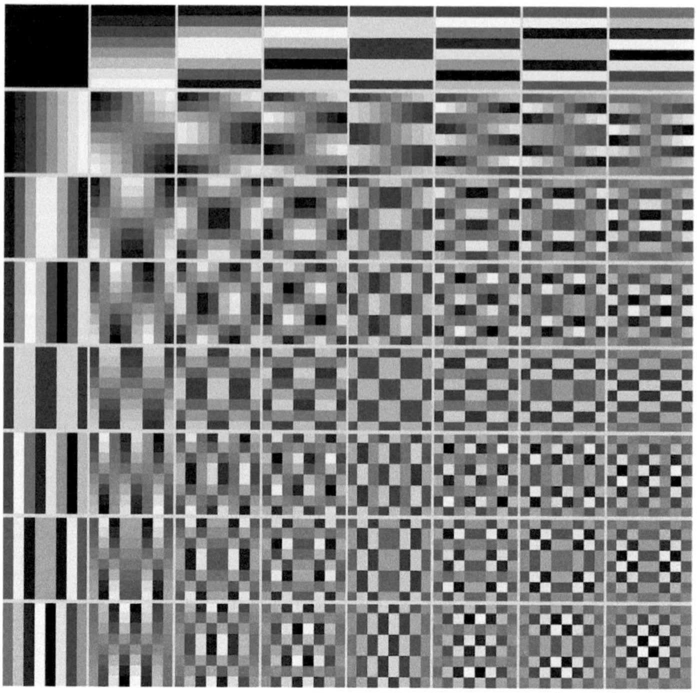

Fig. A.1 Cosine base functions of the two-dimensional Discrete Cosine Transform (DCT) for 8×8 pixel blocks

functions oscillating at different frequencies. These cosine base functions are illustrated in Fig. A.1. From left to right the vertical frequency increases and from top to bottom the horizontal frequencies increase.

The DCT calculates, how well a pixel block can be represented by each of the shown base functions. The corresponding mathematical equation reads as

$$B_{p,q} = \sum_{m=0}^{M-1} \sum_{n=0}^{N-1} x_{m,n} \cdot \cos \frac{\pi(2m+1)p}{2M} \cdot \cos \frac{\pi(2n+1)q}{2N}$$

where $x_{m,n}$ are the pixel information and $B_{p,q}$ the resulting coefficients in the frequency domain. With these coefficients and the corresponding cosine base functions, the image can be reconstructed by the inverse DCT (IDCT).

References

1. G.E. Moore, Cramming more components onto integrated circuits. Electron. Mag. **38**(8), 114–117 (1965)
2. M. Horowitz, E. Alon, D. Patil, S. Naffziger, R. Kumar, K. Bernstein, Scaling, power, and the future of CMOS, in *Proceedings of the IEEE International Electron Devices Meeting (IEDM)* (2005), pp. 7–13
3. N. Ickes, G. Gammie, M.E. Sinangil, R. Rithe, J. Gu, A. Wang, H. Mair, S.R. Datla, B. Rong, S. Honnavara-Prasad, L. Ho, G. Baldwin, D. Buss, A.P. Chandrakasan, U. Ko, A 28 nm 0.6 V low power DSP for mobile applications. IEEE J. Solid-State Circuits **47**(1), 35–46 (2012)
4. T.D. Burd, T.A. Pering, A.J. Stratakos, R.W. Brodersen, A dynamic voltage scaled microprocessor system. IEEE J. Solid-State Circuits **35**(11), 1571–1580 (2000)
5. T. Fischer, F. Anderson, B. Patella, S. Naffziger, A 90nm variable-frequency clock system for a power-managed Itanium-family processor, in *Proceedings of the IEEE International Solid-State Circuits Conference (ISSCC)* (2005), pp. 294–295
6. A. Drake, R. Senger, H. Deogun, G. Carpenter, S. Ghiasi, T. Nguyen, N. James, M. Floyd, V. Pokala, A distributed critical-path timing monitor for a 65nm high-performance microprocessor, in *Proceedings of the IEEE International Solid-State Circuits Conference (ISSCC)* (2007), pp. 398–399
7. K.A. Bowman, J.W. Tschanz, N.S. Kim, J.C. Lee, C.B. Wilkerson, S.-L.L. Lu, T. Karnik, V.K. De, Energy-efficient and metastability-immune timing-error detection and instruction-replay-based recovery circuits for dynamic-variation tolerance, in *Proceedings of the IEEE International Solid-State Circuits Conference (ISSCC)* (2008), pp. 402–403
8. S. Das, C. Tokunaga, S. Pant, W.H. Ma, S. Kalaiselvan, K. Lai, D.M. Bull, D.T. Blaauw, RazorII: In situ error detection and correction for PVT and SER tolerance. IEEE J. Solid-State Circuits **44**(1), 32–48 (2009)
9. D. Bull, S. Das, K. Shivshankar, G. Dasika, K. Flautner, D. Blaauw, A power-efficient 32b ARM ISA processor using timing-error detection and correction for transient-error tolerance and adaptation to PVT variation, in *Proceedings of the IEEE International Solid-State Circuits Conference (ISSCC)* (2010), pp. 284–285
10. D. Ernst, N.S. Kim, S. Das, S. Pant, R. Rao, T. Pham, C. Ziesler, D. Blaauw, T. Austin, K. Flautner, T. Mudge, Razor: A low-power pipeline based on circuit-level timing speculation, in *Proceedings of the 36th Annual IEEE/ACM International Symposium MICRO-36 Microarchitecture* (2003), pp. 7–18
11. J. Tschanz, K. Bowman, S.-L. Lu, P. Aseron, M. Khellah, A. Raychowdhury, B. Geuskens, C. Tokunaga, C. Wilkerson, T. Karnik, V. De, A 45nm resilient and adaptive microprocessor core for dynamic variation tolerance, in *Proceedings of the IEEE International Solid-State Circuits Conference (ISSCC)* (2010), pp. 282–283

M. Wirnshofer, *Variation-Aware Adaptive Voltage Scaling for Digital CMOS Circuits*, Springer Series in Advanced Microelectronics 41, DOI 10.1007/978-94-007-6196-4,

12. The International Technology Roadmap for Semiconductors (ITRS): Design (2011), http://www.itrs.net/Links/2011ITRS/2011Chapters/2011Design.pdf
13. S.K. Saha, Modeling process variability in scaled CMOS technology. IEEE Des. Test Comput. **27**(2), 8–16 (2010)
14. P. Bai, C. Auth, S. Balakrishnan, M. Bost, R. Brain, V. Chikarmane, R. Heussner, M. Hussein, J. Hwang, D. Ingerly, R. James, J. Jeong, C. Kenyon, E. Lee, S.-H. Lee, N. Lindert, M. Liu, Z. Ma, T. Marieb, A. Murthy, R. Nagisetty, S. Natarajan, J. Neirynck, A. Ott, C. Parker, J. Sebastian, R. Shaheed, S. Sivakumar, J. Steigerwald, S. Tyagi, C. Weber, B. Woolery, A. Yeoh, K. Zhang, M. Bohr, A 65nm logic technology featuring 35nm gate lengths, enhanced channel strain, 8 Cu interconnect layers, low-k ILD and 0.57μm^2 SRAM cell, in *Proceedings of the IEEE International Electron Devices Meeting (IEDM)* (2004), pp. 657–660
15. A. Asenov, S. Kaya, A.R. Brown, Intrinsic parameter fluctuations in decananometer MOSFETs introduced by gate line edge roughness. IEEE J. Solid-State Circuits **50**(5), 1254–1260 (2003)
16. C.H. Diaz, H.-J. Tao, Y.-C. Ku, A. Yen, K. Young, An experimentally validated analytical model for gate line-edge roughness (LER) effects on technology scaling. IEEE J. Solid-State Circuits **22**(6), 287–289 (2001)
17. K.J. Kuhn, C. Kenyon, A. Kornfeld, M. Liu, A. Maheshwari, W. Shih, S. Sivakumar, G. Taylor, P. VanDerVoorn, K. Zawadzki, Managing process variation in Intel's 45nm CMOS technology. Intel Technol. J. **12**, 93–109 (2008)
18. A. Asenov, S. Kaya, J.H. Davies, Intrinsic threshold voltage fluctuations in decanano MOSFETs due to local oxide thickness variations. IEEE J. Solid-State Circuits **49**(1), 112–119 (2002)
19. K.J. Kuhn, M.D. Giles, D. Becher, P. Kolar, A. Kornfeld, R. Kotlyar, S.T. Ma, A. Maheshwari, S. Mudanai, Process technology variation. IEEE Trans. Electron Devices **58**(8), 2197–2208 (2011)
20. M.J.M. Pelgrom, A.C.J. Duinmaijer, A.P.G. Welbers, Matching properties of MOS transistors. IEEE J. Solid-State Circuits **24**(5), 1433–1439 (1989)
21. T. Sakurai, A.R. Newton, Alpha-power law MOSFET model and its applications to CMOS inverter delay and other formulas. IEEE J. Solid-State Circuits **25**(2), 584–594 (1990)
22. J. Choi, C.-Y. Cher, H. Franke, H. Hamann, A. Weger, P. Bose, Thermal-aware task scheduling at the system software level, in *Proceedings of the ACM/IEEE International Symposium on Low Power Electronics and Design (ISLPED)* (2007), pp. 213–218
23. S. Borkar, T. Karnik, S. Narendra, J. Tschanz, A. Keshavarzi, V. De, Parameter variations and impact on circuits and microarchitecture, in *Proceedings of the Design Automation Conference (DAC)* (2003), pp. 338–342
24. A. Bravaix, V. Huard, D. Goguenheim, E. Vincent, Hot-carrier to cold-carrier device lifetime modeling with temperature for low power 40nm Si-bulk NMOS and PMOS FETs, in *Proceedings of the IEEE International Electron Devices Meeting (IEDM)* (2011)
25. B. Kaczer, S. Mahato, V.V. de Almeida Camargo, M. Toledano-Luque, P.J. Roussel, T. Grasser, F. Catthoor, P. Dobrovolny, P. Zuber, G. Wirth, G. Groeseneken, Atomistic approach to variability of bias-temperature instability in circuit simulations, in *Proceedings of the IEEE International Reliability Physics Symposium (IRPS)* (2011), pp. 915–919
26. T. Grasser, H. Reisinger, P. Wagner, F. Schanovsky, W. Goes, B. Kaczer, The time dependent defect spectroscopy (TDDS) for the characterization of the bias temperature instability, in *Proceedings of the IEEE International Reliability Physics Symposium (IRPS)* (2010), pp. 16–25
27. H. Reisinger, T. Grasser, K. Hofmann, W. Gustin, C. Schlünder, The impact of recovery on BTI reliability assessments, in *Proceedings of the IEEE International Integrated Reliability Workshop Final Report (IRW)* (2010), pp. 12–16

28. F.R. Chouard, C. Werner, D. Schmitt-Landsiedel, M. Fulde, A test concept for circuit level aging demonstrated by a differential amplifier, in *Proceedings of the IEEE International Reliability Physics Symposium (IRPS)* (2010), pp. 826–829
29. S. Drapatz, K. Hofmann, G. Georgakos, D. Schmitt-Landsiedel, Impact of fast-recovering NBTI degradation on stability of large-scale SRAM arrays, in *Proceedings of the IEEE European Solid-State Device Research Conference (ESSDERC)* (2010), pp. 146–149
30. T. Grasser, P.-J. Wagner, H. Reisinger, T. Aichinger, G. Pobegen, M. Nelhiebel, B. Kaczer, Analytic modeling of the bias temperature instability using capture/emission time maps, in *Proceedings of the IEEE International Electron Devices Meeting (IEDM)* (2011)
31. H. Reisinger, T. Grasser, W. Gustin, C. Schlünder, The statistical analysis of individual defects constituting NBTI and its implications for modeling DC- and AC-stress, in *Proceedings of the IEEE International Reliability Physics Symposium (IRPS)* (2010), pp. 7–15
32. V. Gutnik, A.P. Chandrakasan, Embedded power supply for low-power DSP. IEEE J. Solid-State Circuits **5**(4), 425–435 (1997)
33. S. Gochman, R. Ronen, I. Anati, A. Berkovits, T. Kurts, A. Naveh, A. Saeed, Z. Sperber, R.C. Valentine, The Intel Pentium M processor: Microarchitecture and performance. Intel Technol. J. **7**, 21–36 (2003)
34. Enhanced Intel SpeedStep Technology for the Intel Pentium M Processor (White Paper) (2004), ftp://download.intel.com/design/network/papers/30117401.pdf
35. P. Macken, M. Degrauwe, M. Van Paemel, H. Oguey, A voltage reduction technique for digital systems, in *Proceedings of the IEEE International Solid-State Circuits Conference (ISSCC)* (1990), pp. 238–239
36. D. Lorenz, M. Barke, U. Schlichtmann, Aging analysis at gate and macro cell level, in *Proceedings of the IEEE/ACM International Conference on Computer-Aided Design (ICCAD)* (2010), pp. 77–84
37. V. Zolotov, J. Xiong, H. Fatemi, C. Visweswariah, Statistical path selection for at-speed test. IEEE Trans. Comput.-Aided Des. Integr. Circuits Syst. **29**(5), 749–759 (2010)
38. K.A. Bowman, J.W. Tschanz, N.S. Kim, J.C. Lee, C.B. Wilkerson, S.L.L. Lu, T. Karnik, V.K. De, Energy-efficient and metastability-immune resilient circuits for dynamic variation tolerance. IEEE J. Solid-State Circuits **44**(1), 49–63 (2009)
39. D. Bull, S. Das, K. Shivashankar, G.S. Dasika, K. Flautner, D. Blaauw, A power-efficient 32 bit ARM processor using timing-error detection and correction for transient-error tolerance and adaptation to PVT variation. IEEE J. Solid-State Circuits **46**(1), 18–31 (2011)
40. S. Das, D. Roberts, S. Lee, S. Pant, D. Blaauw, T. Austin, K. Flautner, T. Mudge, A self-tuning DVS processor using delay-error detection and correction. IEEE J. Solid-State Circuits **41**(4), 792–804 (2006)
41. M. Eireiner, S. Henzler, G. Georgakos, J. Berthold, D. Schmitt-Landsiedel, In-situ delay characterization and local supply voltage adjustment for compensation of local parametric variations. IEEE J. Solid-State Circuits **42**(7), 1583–1592 (2007)
42. S. Mitra, Circuit failure prediction for robust system design in scaled CMOS, in *Proceedings of the IEEE International Reliability Physics Symposium (IRPS)* (2008), pp. 524–531
43. M. Wirnshofer, L. Heiss, G. Georgakos, D. Schmitt-Landsiedel, A variation-aware adaptive voltage scaling technique based on in-situ delay monitoring, in *Proceedings of the 14th IEEE International Symposium on Design and Diagnostics of Electronic Circuits and Systems (DDECS)* (2011), pp. 261–266
44. M. Wirnshofer, L. Heiss, G. Georgakos, D. Schmitt-Landsiedel, An energy-efficient supply voltage scheme using in-situ pre-error detection for on-the-fly voltage adaptation to PVT variations, in *Proceedings of the 13th International Symposium on Integrated Circuits (ISIC)* (2011), pp. 94–97
45. M. Wirnshofer, L. Heiss, A.N. Kakade, N. Pour Aryan, G. Georgakos, D. Schmitt-Landsiedel, Adaptive voltage scaling by in-situ delay monitoring for an image processing circuit, in *Proceedings of the 15th IEEE International Symposium on Design and Diagnostics of Electronic Circuits and Systems (DDECS)* (2012), pp. 205–208

46. G.A. Rincon-Mora, P.E. Allen, A low-voltage, low quiescent current, low drop-out regulator. IEEE J. Solid-State Circuits **33**(1), 36–44 (1998)
47. M. Lüders, B. Eversmann, J. Gerber, K. Huber, R. Kuhn, D. Schmitt-Landsiedel, R. Brederlow, A fully-integrated system power aware LDO for energy harvesting applications, in *Proceedings of the Symposium on VLSI Circuits (VLSI)* (2011), pp. 244–245
48. N. Pour Aryan, L. Heiss, D. Schmitt-Landsiedel, G. Georgakos, M. Wirnshofer, Comparison of in-situ delay monitors for use in adaptive voltage scaling. Adv. Radio Sci. **10** (2012)
49. I. Jung, G. Jung, J. Song, M.-Y. Kim, J. Park, S.B. Park, C. Kim, A 0.004-mm^2 portable multiphase clock generator tile for 1.2-GHz RISC microprocessor, in *Proceedings of the Symposium on VLSI Circuits (VLSI)* (2006), pp. 104–105
50. Y.-J. Wang, S.-K. Kao, S.-I. Liu, All-digital delay-locked loop/pulsewidth-control loop with adjustable duty cycles. IEEE J. Solid-State Circuits **41**(6), 1262–1274 (2006)
51. S. Tam, R.D. Limaye, U.N. Desai, Clock generation and distribution for the 130-nm Itanium 2 processor with 6-MB on-die L3 cache. IEEE J. Solid-State Circuits **39**(4), 636–642 (2004)
52. S. Rusu, S. Tam, H. Muljono, J. Stinson, D. Ayers, J. Chang, R. Varada, M. Ratta, S. Kottapalli, S. Vora, A 45 nm 8-core enterprise Xeon processor. IEEE J. Solid-State Circuits **45**(1), 7–14 (2010)
53. R.B. Hitchcock, G.L. Smith, D.D. Cheng, Timing analysis of computer hardware. IBM J. Res. Dev. **26**(1), 100–105 (1982)
54. J. Bhasker, C. Rakesh, *Static Timing Analysis for Nanometer Designs: A Practical Approach* (Springer, New York, 2009)
55. C. Knoth, U. Schlichtmann, Characterization of standard cells, in *Process Variations and Probabilistic Integrated Circuit Design* (Springer, New York, 2012)
56. A. Markov, Rasprostranenie zakona bol'shih chisel na velichiny, zavisyaschie drug ot druga, in *Izvestiya Fiziko-Matematicheskogo Obschestva pri Kazanskom Universitete*, vol. 15 (1906), pp. 135–156. Reprinted in: R. Howard, *Dynamic Probabilistic Systems: Markov Models*, Dover Books on Mathematics, Appendix B (1971)
57. C. Burge, S. Karlin, Prediction of complete gene structures in human genomic DNA. J. Mol. Biol. **268**(1), 78–94 (1997)
58. J. Sebat, B. Lakshmi, J. Troge, J. Alexander, J. Young, P. Lundin, S. Maner, H. Massa, M. Walker, M. Chi, N. Navin, R. Lucito, J. Healy, J. Hicks, K. Ye, A. Reiner, T.C. Gilliam, B. Trask, N. Patterson, A. Zetterberg, M. Wigler, Large-scale copy number polymorphism in the human genome. Science **305**(5683), 525–528 (2004)
59. C.E. Shannon, A mathematical theory of communication. Bell Syst. Tech. J. **27**, 623–656 (1948)
60. L. Page, S. Brin, R. Motwani, T. Winograd, The PageRank citation ranking: Bringing order to the web. Stanford InfoLab, Tech. Rep. (1999)
61. R.A. Jarrow, D. Lando, S.M. Turnbull, A Markov model for the term structure of credit risk spreads. Rev. Financ. Stud. **10**, 481–523 (1997)
62. K. McAlpine, E. Miranda, S. Hoggar, Making music with algorithms: A case-study system. Comput. Music J. **23**, 19–30 (1999)
63. F. Pachet, P. Roy, G. Barbieri, Finite-length Markov processes with constraints, in *Proceedings of the 22nd International Joint Conference on Artificial Intelligence (IJCAI)* (2011), pp. 635–642
64. A. Papoulis, S.U. Pillai, *Probability, Random Variables, and Stochastic Processes* (McGraw-Hill, New York, 2002)
65. N. Weste, D. Harris, *CMOS VLSI Design: A Circuits and Systems Perspective* (Pearson Education, Upper Saddle River, 2005)
66. Y. Taur, T. Ning, *Fundamentals of Modern VLSI Devices* (Cambridge University Press, Cambridge, 2009)
67. M. El-Nozahi, A. Amer, J. Torres, K. Entesari, E. Sanchez-Sinencio, High PSR low drop-out regulator with feed-forward ripple cancellation technique. IEEE J. Solid-State Circuits **45**(3), 565–577 (2010)

68. S.D. Poisson, *Recherches sur la probabilité des jugements en matière criminelle et en matière civile: Précédées des règles générales du calcul des probabilités* (Bachelier, Paris, 1837). Reprinted by Adamant Media Corporation (2001)
69. G.K. Wallace, The JPEG still picture compression standard. IEEE Trans. Consum. Electron. **38**(1), 18–34 (1992)
70. M.A. Tekalp, *Digital Video Processing* (Prentice-Hall, New York, 1995)
71. L.J. D'Luna, W.A. Cook, R.M. Guidash, G.W. Brown, T.J. Tredwell, J.R. Fischer, T. Tarn, An 8×8 discrete cosine transform chip with pixel rate clocks, in *Proceedings of the Third Annual IEEE ASIC Seminar and Exhibit* (1990)
72. M. Ashouei, J. Hulzink, M. Konijnenburg, J. Zhou, F. Duarte, A. Breeschoten, J. Huisken, J. Stuyt, H. de Groot, F. Barat, J. David, J. Van Ginderdeuren, A voltage-scalable biomedical signal processor running ECG using 13pJ/cycle at 1MHz and 0.4V, in *Proceedings of the IEEE International Solid-State Circuits Conference (ISSCC)* (2011), pp. 332–334
73. J. Kwong, Y. Ramadass, N. Verma, M. Koesler, K. Huber, H. Moormann, A. Chandrakasan, A 65nm sub-Vt microcontroller with integrated SRAM and switched-capacitor DC-DC converter, in *Proceedings of the IEEE International Solid-State Circuits Conference (ISSCC)* (2008), pp. 318–319
74. N. Lotze, Y. Manoli, A 62mV 0.13um CMOS standard-cell-based design technique using Schmitt-Trigger logic, in *Proceedings of the IEEE International Solid-State Circuits Conference (ISSCC)* (2011), pp. 340–342

MIX
Papier aus verantwortungsvollen Quellen
Paper from responsible sources
FSC® C105338

If you have any concerns about our products,
you can contact us on
ProductSafety@springernature.com

In case Publisher is established outside the EU,
the EU authorized representative is:
Springer Nature Customer Service Center GmbH
Europaplatz 3, 69115 Heidelberg, Germany

Printed by Libri Plureos GmbH
in Hamburg, Germany